owning the future

BOOKS BY

SETH SHULMAN

The Threat at Home: Confronting the Toxic Legacy of the U.S. Military

Owning the Future

Owning the Future

Seth Shulman

Houghton Mifflin Company

BOSTON NEW YORK

1999

Library of Congress Cataloging-in-Publication Data

Shulman, Seth.
Owning the future / Seth Shulman.
p. cm.
Includes index.
ISBN 0-395-84175-5
1. Technological innovation — Economic aspects — Forecasting.
2. Diffusion of innovations — Economic aspects—Forecasting.
3. Information technology — Economic aspects — Forecasting.
4. High technology industries — Forecasting. 5. Knowledge management.
6. Intellectual property. 7. Patent licenses. 8. Exchanges of patents and technical information. I. Title.
HC79.T4S53 1999
338'.064 — dc21 98-39653 CIP

Book design by Anne Chalmers
Typefaces: Minion, Triplex Serif Bold

Printed in the United States of America
QUM 10 9 8 7 6 5 4 3 2 1

FOR

LAURA

The preservation of the means of knowledge among the lowest ranks is of more importance to the public than all the property of all the rich men in the country.

— John Adams (1735–1826)

Contents

Preface

The unlikely starting point for this book about emerging technological developments is an old bookshelf in the town of Franklin, Massachusetts.

The bookshelf is modest by any measure. Four shelves in a glass-fronted case hold 116 musty, well-worn books. Many of the titles are illegible. There is a book on logic, several on Christian history, the legal works of William Blackstone, four volumes of philosophical essays by John Locke. The books are a gift from Benjamin Franklin, the town's namesake. All but forgotten even by many Franklin residents, these four shelves hold the nation's first public library.

My research for this book took me to many offbeat locales. I toured the steely, modernist sprawl of Crystal City, Virginia, home of the U.S. Patent and Trademark Office. I thumbed through some of Thomas Edison's five hundred extraordinary, and rarely viewed, laboratory notebooks in an underground vault beneath the complex where he worked in West Orange, New Jersey. But of all the spots I visited, none seems more pertinent than the bookshelf in Franklin.

I traveled to the town because I intended to write about the threat posed by the expanding private ownership of vast electronic databases. I was curious about the extent to which the pay-per-view control of the so-called information industry—with its electronic troves of news stories, business information, government documents, and court decisions—was already clashing with the age-old "free-to-all" tradition of public libraries. What better touchstone could there be than a firsthand visit to the nation's oldest public library?

When I first saw it, however, the library filled me with dismay. I realized at once that the imposing granite edifice, reminiscent of an oversized bank, was not old enough to have housed the nation's first public library. Annoyed at myself for not finding out more about Franklin's library over the phone, I feared my pilgrimage would yield little more than an "On this site stood. . ." commemorative plaque.

In this state of disappointment, thinking that I had foolishly wasted a trip, I approached the circulation desk. There on my right I noticed the modest bookcase I had unwittingly come to see. I realized how little I knew of the history of the public library, an institution I take largely for granted.

As it turns out, I was not alone in my ignorance of Franklin's historic book collection. I was amazed to discover that the harried librarians, busily checking out books on space exploration and the latest children's videos, didn't know much about the chunk of history in their midst either. Of course, they were aware of the pathbreaking collection, but only in the way townspeople might be familiar with a big equestrian statue in the park that they passed every day without ever learning who it commemorates. Clearly, the Franklin bookshelf was not something many people ever asked about.

After I enlisted the help of all three on-duty librarians, however, one finally exhumed some old town documents from a closed shelf in a back room. I happily spent the rest of the morning piecing together the largely unheralded tale, drawing especially from a copious volume of town history published in 1878.

I found myself captivated by the unexpected story of the books, which were something of a fluke. In 1778 the area's enterprising residents had sought to flatter the eminent Benjamin Franklin by naming their newly incorporated town for him. In return for the honor, they asked Franklin to donate a bell for the steeple of their town meetinghouse. Instead he donated books, and a letter counseling the townsfolk that "sense is preferable to sound."

But the townspeople didn't know exactly what to do with the books. For years the collection shuttled from one location to another, from the foyer of the minister's house to the attic of a private home across from the town cemetery. For a while the books were even warehoused in a local barn.

In no fewer than ten town meetings, Franklin residents debated who should be allowed access to the collection. A vocal faction, led by Nathaniel Emmons, the town's Congregational minister (and de facto first librarian) argued that the books should be lent only to members of the parish. The issue was so contested that the town's residents even wrote to Benjamin Franklin for guidance in the matter, but if he sent a reply, no record of it survives.

Finally, in December 1790, nearly five years after the books were donated, a majority of town residents voted to overrule Emmons and make them available to all the town's citizens. It was through this small, parochial decision that the progenitor of the nation's public library system was born.

In the months since my visit to Franklin, I have thought often about the unprecedented decision reached by this local group. Of course, Franklin's collection is paltry compared with the robust tradition of libraries dating to antiquity. As early as 300 B.C., the Greek empire's famous library in Alexandria, Egypt, reportedly held some 500,000 scrolls. Substantial book collections in Europe's monasteries date to the Middle Ages, if not earlier. Even by the comparatively recent standards of the United States, scores of private libraries predate Franklin's collection. Harvard College's book collection was more than twice the size of Franklin's even when it was originally bequeathed in 1638. And Ben Franklin himself had started a membership library in Philadelphia as early as 1731.

Yet never before in this long sweep of history had any governing body in charge of a collection of books chosen to share it with so broad and generous a conception of the public. Never before had a bastion of accumulated written knowledge, however modest, been subject to so democratic an interpretation of open access. Inspired by the same Enlightenment sensibilities that had fueled the drafting of the U.S. Constitution just four years earlier, the decision of Franklin's residents to freely share the town's storehouse of knowledge helped transform civil society. From this humble bookshelf, the public library would eventually sprout today's 16,000 local branches throughout the nation.

Decades after the town's fateful resolution, Franklin native Horace Mann—often called the founder of public education in the United States—would credit the books he read in the town's original collection as an inspiration for his career in education. Mann criticized some of Benjamin Franklin's selections for glorifying war and armed conflicts. But with few books in his own home as a boy, Mann had only praise for the nation's earliest public library and the world of ideas it opened to him. In his own career, Mann championed the establishment of free public libraries in every school district.

As I beheld Franklin's bookcase, I marveled at the inspired, principled stance these New England residents had taken more than two centuries ago. However serendipitous their story, and however modest the scale of the endeavor, Franklin's citizens had resolved at a key juncture that the world's knowledge should be accessible to all. That outcome was not inevitable; it resulted from inspired and lengthy debate. Yet, ignoring convention and historical precedent, the townspeople of Franklin upheld an ideal that still merits our attention today.

Unfortunately, the history of the town's deliberations is not the only piece of the story that has been largely forgotten. Today, despite our democratic tradition, we seem to have forgotten Franklin's egalitarian ideals as well.

As a journalist chronicling technological developments for the past fifteen years, I have grown increasingly troubled by the steady barrage of fights over intellectual property. As computer and software companies argue about possession of exclusive rights to conceptual breakthroughs, and university researchers bicker over ownership of the latest biotechnology process, something seems gravely amiss. Lost in the news stories about who wins and who loses these legal battles is the bigger picture of the public's dwindling access and claim to the knowledge and techniques upon which our economy increasingly depends.

All too often today, private parties are quietly auctioning off formerly public assets to the highest bidder, eroding our public institutions and even our shared heritage. The shift is often subtle and frequently passes unnoticed. But by passively acquiescing to the expansion of private ownership into heretofore public realms, we are losing something precious.

Covering science and technology for many publications, including *Nature, Time, Discover, Bioscience,* and *Technology Review,* I have become convinced that the emerging battles over knowledge assets are increasingly vital to each of us. The disputes often begin in the rarefied atmosphere of an academic laboratory or a sequestered office in a software firm. But their effects are pervasive, influencing our jobs, our schools, and our relations with colleagues and neighbors. The results of these battles shape the type of medical care we receive and the availability of the food in our refrigerators. Battles over ownership of the rights to the world's storehouses of knowledge already have largely determined who is rich and who is poor in our society. On a global scale, these battles have begun to foster international conflicts, a trend that is sure to grow in coming years.

This book offers a number of dispatches from the battlefronts, chronicling the deep and disturbing fights taking place in a variety of fields over rights to knowledge. My aim in these pages is to document the extent to which we have expanded our notion of ownership into new and uncharted terrain, and to help explain what's at stake. Along the way, as I have tried to sort out the technical details and legal nuances in these stories, I have repeatedly relied on a sense of democratic first principles about access to and control over knowledge and information. And here I have drawn my inspiration from the residents of Franklin, whose principled stance proved both sound and timeless.

Part I

Knowledge Wars

I

Gold Rush in the Idea Economy

IN THE WANING YEARS of the twentieth century, in an era of unprecedented technological development and instantaneous telecommunications, freely shared knowledge is becoming an endangered species. In the United States, politicians and pundits spout lofty platitudes about public education and public access to information as the bedrock of democratic society, but reality belies the rhetoric. As ideas, concepts, blueprints, and codes become the most sought-after commodities in our new knowledge-based economy, people are hoarding, fighting over, and seeking to control them as never before. As one intellectual property lawyer put it, today's gold rush on knowledge assets "makes the monopolies of the nineteenth-century robber barons look like penny-ante operations."

Today doctors are claiming to own the medical procedures they once shared openly with colleagues. Software firms are winning monopolies on the basic building blocks of computer code needed to write new programs and using their ownership to stymie would-be competitors. Scientists at the nation's top universities and research institutes complain that collegial discourse has withered in the face of proprietary claims and secrecy among researchers. Drug companies are systematically gathering wild plants, insects, and microorganisms from the globe's far reaches and claiming exclusive dominion over the chemicals they contain. Even our own genetic makeup is being sold: of the portion of the human genome that has been mapped, roughly a third is already privately owned.

Without thoughtful intervention, the current trajectory promises nothing less than an uncontrolled stampede to auction off our technological and cultural heritage, a future of increasing conflict and dissension, and the specter of an ominous descent into a new Dark Age.

To envision where we're headed, imagine a city in which expensive metered taxicabs offer the only means of transportation. Now imagine further that instead of a robust system of public thoroughfares, the city offers only a labyrinthine collection of separate, privately owned roads, pockmarked by barricades, guarded checkpoints, and costly tolls. People must spend a large portion of their income and even more of their time getting from one place to another. Face-to-face gatherings of people from disparate parts of town are rare; casual, spontaneous meetings, almost nonexistent. Not only does commerce suffer, its shape disproportionally caters to the rare few who can afford to navigate effectively. Such a world differs so radically from the freedom of movement we take for granted that it is almost hard to picture. Taxis and toll roads can certainly be useful tools as part of a diverse transportation system. But they cannot function successfully in the absence of a shared infrastructure.

The same holds true of the domain commonly known as intellectual property. Ideas, formulas, and concepts need to move, just as people do, and the image of a subdivided and barricaded world captures many aspects of the emerging knowledge economy unless we intervene quickly to change it.

Already we see an unprecedented profusion of intellectual property litigation. Hospitals are battling with researchers over rights to control new treatments, universities haul faculty members into court to establish who will profit from their research, and genetic engineering firms wage seemingly endless fights over proprietary rights to techniques and materials. Shakespearean scholars have even gone to court over who owns rights to a particular interpretation of *Hamlet*.

The bards of the new economy tell us repeatedly that knowledge is now a corporation's most valuable resource. "The most important form of property is now intangible. It is super-symbolic. It is knowledge," write high-tech gurus Alvin and Heidi Toffler, for instance. "Knowledge becomes the ultimate substitute — the central resource of an advanced economy," they contend, "because it reduces the need for raw materials, labor, time, space, capital and other inputs."

As Alvin Toffler notes, the knowledge economy was heralded in 1956, the year blue-collar manufacturing workers ceased to be a majority of the U.S. workforce, surpassed by white-collar, service-oriented employees. In a parallel progression, given the structure of what *Newsweek* has called the "New Economy of Ideas," we are moving steadily toward a situation in which many companies will earn the bulk of their profits not as manufacturers but as titleholders and gatekeepers to fiefdoms of knowledge. Over the past several years the electronics firm Texas Instruments has earned a larger portion of its $200 million in annual profits from licensing patents and winning infringement cases than from selling products. With scores of patents covering many of the basic techniques for manufacturing computer chips, Texas Instruments is in a strong position to pull off such a feat. But in the current economy, the firm must be seen as a harbinger of things to come.

What happens when businesses, governmental agencies, and academic institutions earn the majority of their profits from their intellectual property holdings?

Imagine you are dying of cancer and could be aided by a life-saving, FDA-approved cancer treatment developed by a new drug company. But your doctor can't give it to you because a rival pharmaceutical firm — which has no comparable product of its own — wins a court order to halt the sale of the treatment, claiming to own a segment of its underlying technology.

Imagine you are a university researcher who makes a lucrative discovery during long hours in the lab. But a dispute over ownership rights turns so bitter that you wind up in prison after the university successfully presses criminal charges, alleging that your ideas are in fact the university's stolen intellectual property.

Imagine you are a farmer who, after harvesting your crop, learns that it is not entirely yours: the agrochemical firm that sold you the seeds says you can't replant a portion of your harvest as seed for next year's crop, as your family has done for generations, because the company owns the crop's germplasm — its blueprint — even in successive generations.

Do these scenarios seem farfetched and extreme? *All of them are drawn from legal cases that have already occurred.* Nor can any of them be easily dismissed as an anomaly; rather, as a tiny sampling of the skirmishes breaking out regularly on the frontlines of the knowledge wars, they suggest an emerging norm.

The fact is, in almost every high-tech field from software design to agribusiness, the push to own the rights to information, concepts, and techniques raises a host of difficult questions. Who will have — or lose — access to the world's knowledge and at what cost? Is owning rights to a concept really akin to owning a piece of real estate, as the phrase "intellectual property" implies? Are there some types of know-how that shouldn't be bought and sold?

Questions like these have long occupied philosophers and legal scholars, but they have so far been largely absent from public debate. Today, however, none of us can afford to remain uninformed or silent. The challenge before us all, as participants in a democracy, is to lay bare the fundamental fallacy of a system that lavishly rewards the incremental innovations of individuals but ignores our collective stake in society's accumulated wealth of know-how. Rectifying the situation will not be easy. It will require nothing short of remaking the civic sphere, deciding which pieces of our intellectual and cultural heritage should be collectively preserved and even subsidized as part of the public domain. But the stakes match the immensity of the challenge: nothing less than the integrity of our shared civic institutions rests on the outcome.

Where Do We Begin?

The first issue to examine is the expansion of the notion of ownership into the conceptual realm. In theory, patents and copyrights were designed to give people special rights, not to ideas as such but to their practical applications (in the case of patents) and to their particular expression (as covered by copyright). In the past only tangible innovations could be patented. Up until 1870 the U.S. Patent Office even required patent applicants to submit physical models of their inventions. But twice in the 1800s, the agency's grand and unruly gallery of models burned to the ground and, partly to avoid the costs of storing them, Congress finally dropped that requirement in 1870. That change was a loss for the public. For one thing, the models, housed in Washington, D.C., attracted up to ten thousand visitors per month, bespeaking a degree of public interest and involvement in the patent process rarely matched since. Equally important, the requirement rooted the patent system to reality, by reflecting the standard that a patent should cover *some thing*— either a machine or the embodiment of a particular process — that yields a material result.

Today the patent system has moved so far away from this concrete interpretation of intellectual property that the notion of requiring a model is almost unthinkable. More and more patents cover what some have termed "actionable knowledge," such as the composition of a gene that might be useful in a medical therapy or a programming technique that facilitates a variety of software applications. As Wallace Judd, a computer programmer and president of California-based Mentrix Corporation, has noted, rather than protecting a particular innovation, the current system often authorizes exclusive control of a broad concept. The difference, Judd says, is between a patent on a particular improved mousetrap and a monopoly on the idea of trapping mice. In practice, the distinction Judd notes can be surprisingly subtle. Unfortunately, especially in a knowledge-based economy, it is exceedingly difficult to distinguish between the two types of claims.

The implications of the change are profound. The benefit of allowing someone to protect their particular innovation in mousetrap design is clear: it offers an incentive for people to bring new inventions to the market and allows them to more easily recoup their research costs, safe in the knowledge that their hard-earned design won't be stolen. Patent Office representative Charles Van Horn explains that the power of patents derives from their unique ability to allow the bearer "the crucial right to exclude others." This protection, he says, "enables a person who is licensed by a patent holder to invest the dollars often necessary to bring an idea to the public." Van Horn's point is arguable, though it seems plausible enough when applied to specific innovations. But most people would agree that conferring exclusive rights to the concept of trapping mice gives the lucky patent owner an unfair advantage and establishes a new tollbooth where there needn't be one.

A growing number of scientists, business executives, and even patent lawyers worry that the system has gone awry. Many argue that our patent system has not adapted to the radically new kinds of claims now being foisted upon it. It doesn't help matters that the system is vested in age-old ideas about land ownership, as indicated by the term "intellectual *property*," as well as by the way the Patent Office files ownership claims to this highly conceptual landscape. Even though the quantity of paper at the U.S. Patent Office has grown phenomenally, the design of the system hasn't changed since Thomas Jefferson, the nation's first patent administrator, devised it. In a rabbit's warren of a library, the stapled sheaves of paper establishing each individual patent are stacked chronologically in piles of precisely defined categories within "shoebox" file drawers. Hundreds of rows of these drawers cover aisle after aisle from floor to ceiling in a system that emulates the way a registry of deeds accounts for parcels of land. Yet this antiquated, property-based system must now cope with what John Perry Barlow, former Grateful Dead lyricist cum commentator on cyberspace, has nicely termed "the most unreal estate imaginable."

In the United States today, the courts are conflicted about whether you own the rights to your own telephone number, your genetic material, or even your life story. But what does seem largely settled is that all this unreal estate can legitimately be traded, bought, and sold. From U.S. Supreme Court rulings to international trade negotiations, the industrialized world has moved surprisingly swiftly to institutionalize the notion of knowledge as a commodity. Unfortunately for most of us, many of the key decisions that have solidified this expansive view of ownership have come about with little public debate. Because these new realms are still so poorly understood, and we lack a workable framework to effectively delimit the scope of intellectual property claims, it seems that anything can be owned today.

Part of the problem reflects Judd's mousetrap complaint: the difficulty, especially in high-tech fields, of drawing a line between a specific invention and an entire area of endeavor.

When the Wisconsin-based biotechnology firm Agracetus succeeded in inserting particular genes into cotton in 1988, it sought patent protection not for its modified cotton plant or its novel process, but for *all genetically engineered cotton* achieved by any means. The U.S. Patent Office initially said yes, then changed its mind after a chorus of objections. The ensuing legal battle has yet to be fully resolved.

When a team of researchers at the National Institutes of Health (NIH) made medical history by employing a new technique to successfully treat two girls with a rare genetic disorder, they parlayed the experiment into a patent on *all* so-called ex vivo human gene therapy. The claim is so vast that doctors around the country watched dumbstruck as exclusive rights to an entire new field of medicine were sold to the highest bidder, ultimately, in this case, a Swiss pharmaceutical firm.

Or take a case straight out of *Jurassic Park*. The premise of Michael Crichton's book and the subsequent blockbuster movie was that mosquitoes, hermetically encased in amber for millennia, contained in their bellies the DNA of dinosaurs they had bitten — DNA that could be used to clone the Jurassic creatures. Crichton made a fortune by bringing the idea to the public, but in fact a private titleholder owns any conceivable application that might derive from the concept of using ancient DNA encased in amber. In a well-publicized and controversial experiment in 1995, Raul J. Cano, a prominent molecular biologist and founder of the San Francisco–based Ambergen Corporation, claimed to have successfully extracted DNA from an ancient microorganism trapped in amber. Many scientists expressed skepticism, speculating that the purportedly ancient genes Cano found might have derived from modern-day microorganisms contaminating his samples. But the skeptics are not allowed to freely replicate the experiment, as scientific tradition would dictate. Based on his single experiment, Cano garnered a patent that gives him two decades of exclusive rights: not to the microorganism he found or to the method he used to extract it, but to the recovery of any and all organisms — bacteria, fungi, protozoa, viruses, microalgae, pollen, or arthropods — from amber or other natural resins. Even Albert P. Halluin, Cano's patent lawyer, concedes that the monopoly ownership Cano won is "extremely broad."

The U.S. patent system, like its European forebears, was designed to reward innovation, guard against secrecy, and, as Abraham Lincoln once put it, to add "the fuel of interest to the fire of genius." The idea is that by offering an inventor monopoly protection for individual inventions, the government can help spur innovation. The notion has, by any estimation, been a powerful and fruitful one. Many of today's patents, however, make a mockery of these idealized goals. Even *Forbes* magazine, not known for its radical critiques of capitalism, decried the situation in an editorial in 1993. Especially in high-tech fields, *Forbes* editors wrote, "the patent system has become a lottery in which one lucky inventor gets sweeping rights to a whole class of inventions, and stymies development by inventors."

But the problem goes well beyond the issuance of unfathomably broad patents. Take the case of Ashleigh Brilliant, a California-based writer and former history professor who, notwithstanding his last name, makes his living largely by coining prosaic sayings and claiming private ownership of them. So far, Brilliant has copyrighted more than 7,500 aphorisms. Former television newscaster David Brinkley learned about Brilliant's literary legerdemain the hard way. When Brinkley published his 1996 memoir, *Everybody Is Entitled to My Opinion,* Brilliant sued, documenting that he had copyrighted that saying years before. Brinkley called it "a shakedown," but his publisher, Random House, quietly coughed up royalties for use of the expression. The case was but one of more than a hundred infringement cases Brilliant has successfully mounted.

Corporations, of course, have long registered their pithy slogans and jingles as company trademarks and, like Brilliant, threatened lawsuits to prevent others from using them. But today these kinds of ownership claims have ballooned into unprecedented realms. The Chicago-based Qualitex Company successfully claimed exclusive rights to a color: the "special shade of green-gold" of pads they manufacture for dry-cleaning presses. In a case that reached the U.S. Supreme Court in 1995, Qualitex blocked a rival manufacturer from making a green-gold pad by claiming to own that particular hue as a company trademark. The high court ruled that Qualitex had a valid claim to the color. How could they rule otherwise, Justice Stephen G. Breyer asked in the decision, considering that the courts had already allowed companies to own sounds (such as those three chimes NBC uses with its peacock logo) and even fragrances (in a case allowing a company exclusive use of a scent on sewing thread)?

The examples seem fanciful, but the climate is insidious. The Mattel Corporation, for instance, has hauled a publishing firm into court for producing a hobby magazine for Barbie enthusiasts. Mattel argues that its ownership claims on Barbie extend to the realm of publications about the doll. The esteemed British mathematician Sir Roger Penrose claims rights to a nonrepeating geometric pattern he discovered. Sir Penrose is suing Kimberly-Clark for putting the pattern on its Kleenex brand of quilted toilet paper. One Manhattan law firm specializing in intellectual property law is now even arguing that professional athletes may have the legal right to patent or copyright unique moves that they use in competition, thereby, at least theoretically, preventing their opponents from using them.

Is there no limit to what people or companies can own? Should there be? Several recent court cases have come up against this question when firms have claimed rights to their employees' thoughts. In a case now on appeal, a Texas court ordered Evan Brown, a computer programmer, to reveal his idea for a software procedure to his former employer, DSC Communications of Plano, Texas, even though Brown had never used or developed the notion during his tenure with the firm.

Brown's case grew out of a bitter severance. Brown had been fired after working at DSC for ten years. But before leaving the company, he had mentioned to higher-ups his idea for a program that could automatically convert old software code into newer language. DSC has — so far successfully — claimed rights to this information, arguing that it is covered in an employment agreement Brown signed. The provision in question claims that all ideas an employee might have that relate to DSC's line of business are the rightful property of the company. The company's position, as its general counsel, George Bunt, explains, is, "If a janitor came up with a method of cleaning a hardwood floor suggested to him by his work in cleaning a DSC hardwood floor, technically the idea belongs to DSC."

These kinds of far-reaching claims to intellectual property contribute directly to a stifled research environment. American University law professor James Boyle, one of a new generation of intellectual property specialists to examine the issue, warns, "At some point, the public domain will be so diminished that future creators will be prevented from creating because they won't be able to afford the raw materials they need. An intellectual property system has to insure that the fertile public domain is not converted into a fallow landscape of walled private plots."

Few realize the extent to which we risk just such a fallow landscape. To get a sense of this problem, let's say that you develop a new razor and want to bring it to market. Your chances of success are extremely slim, and not just because a few large firms dominate the production and distribution of razors. To an astonishing extent, these large firms own the useful know-how amassed to date about shaving. No fewer than seventeen patents are listed on the package of Gillette's nonelectric sensor razor. If you, undeterred by the patent portfolios of Gillette and other firms, decide to investigate the matter further at the U.S. Patent Office library, you would face a baffling maze of narrow subdivisions, all laid out and designated like so many reserved parking spaces. Within the large and mature patent category covering cutlery, called Class 30, for instance, section 32 covers "razors with waste-collecting, razor-cleaning and/or dispensing"; section 43.1 covers "razor adjusting means." If your razor happened to tilt with respect to the handle, it would likely infringe U.S. Patent No. 4,026,016 — one of the Gillette razor's seventeen — that stakes a claim to a particular design for a "Razor Blade Assembly" allowing for a tilting blade.

What happens, though, when such proliferating mini-monopolies infiltrate more conceptual realms? Supposedly the Patent Office exempts scientific laws from consideration, but in 1995 Richard Stallman, a computer programmer and well-known critic of the patent system, testified that, in an effort to test the system, a colleague of his succeeded in winning a patent on Kirchoff's Law — an 1845 scientific theory holding that the electric current flowing into a junction equals the current flowing out. In 1994 Roger Schlafly, a California mathematician, even successfully staked a patent claim over two large prime numbers that can be used together as part of a program to encrypt electronic mail.

Technically the patent gives Schlafly the legal right to sue anybody in the United States for using his numbers without permission. "I was kind of interested in pushing the system to see how far you could go with allowable claims," Schlafly told a reporter. But, as Schlafly noted, the numbers do satisfy the U.S. Patent Office's age-old conditions for patentability: they are useful, have not been employed similarly before, and their inclusion in this encryption technique is not obvious to a practitioner in the field.

When she learned of the Schlafly prime-number patent, law professor Pamela Samuelson (now at the University of California at Berkeley), an expert on software patents and copyrights, called it "outrageous," arguing that Schlafly's claims expanded the patent system far beyond legal precedent. But Samuelson couldn't say she was surprised. She had warned, back in 1989, of an impending "crisis in intellectual property law" as an "explosive growth of new technologies" stretches our system of intellectual property into ill-fitting new realms.

In the United States, Congress, state and federal courts, and the largest corporations have begun to fashion the institutional framework for a pervasive, expanded definition of ownership covering everything from the languages computers speak to natural functions of the human body. But for all these bureaucratic changes, few have addressed the vital issue of how society's valuable knowledge assets are now being apportioned. Left unchecked, the rush to claim nuggets of the knowledge economy threatens to limit our choice and diversity, stifle innovation, and foster needless monopolies that will increase the cost of goods and services we now take for granted. Most importantly, it threatens to widen even further the disparity between the haves and the have-nots.

Information and knowledge are not only the preferred new currency of commerce, they are also the enduring lifeblood of our civic culture. In the rush to claim terrain, most players in the new knowledge economy are ignoring this basic fact. An economy of ideas that is not tempered by a sense of civic-minded purpose is a disaster waiting to happen. Even a cursory survey finds administrators of libraries, museums, hospitals, photographic archives, universities, and research institutes all confronting unprecedented commercial pressures from companies and individuals staking private claims to material formerly considered part of the public domain. In each case the private claims threaten the viability of these organizations' public missions.

2

The New Wealth of Nations

Over the past two decades the United States has led the world in significantly expanding the definition of allowable intellectual property claims. A key step occurred with the 1980 U.S. Supreme Court decision that sanctioned the world's first patent on an altered life form. In a five-to-four ruling, the Court determined that a bacterium genetically engineered to "eat" crude oil was in fact a product of human invention. In the famous decision, the court ruled, "Anything under the sun made by man" was fair game for the patent system.

If today's notion of permissible intellectual property claims has broadened, the effect of the change has been greatly enlarged by the central role of technological know-how in our society. Knowledge has been used as a tool by those in power for as long as monarchs, pharaohs, and emperors have consulted their advisers. But never before have knowledge assets so dominated the marketplace.

Just think of the rich technological harvest of the latter half of the twentieth century alone. We've harnessed atomic energy, gone to the moon, deciphered the genetic code, and wired the world for instantaneous telecommunications. And, as many have noted, the pace of technological change seems to be speeding up all the time. We still rightly laud Thomas Edison for his unrivaled lifetime accumulation of 1,093 patents, for instance. But many companies today garner nearly that number of patents *each year.* In 1997, IBM was granted 1,724 U.S. patents. Its empire of active ownership claims numbers well into the tens of thousands and already earns the company close to $1 billion annually.

If you doubt the towering preeminence of knowledge assets in the marketplace, consider the source of the wealth of the richest person alive: the CEO of Microsoft, Bill Gates, made his fortune not by producing a new widget we couldn't live without but by owning a language for computers to speak. Or reflect on the storied initial public offering of the Internet software firm Netscape Communications. Within a week of going public in August 1995, Netscape had made a billionaire of its cofounder and majority owner, Jim Clark. The company had fewer than one hundred employees. It had no profits. What it did have, in its Netscape Navigator software, was ownership rights to an important and popular Internet tool.

As of this writing, Netscape has learned that technological innovation can be a fickle business; the company has found it difficult to sustain the explosive growth it enjoyed upon entering the market. Nonetheless, success stories like Netscape's initial offering feed a dizzying and often confusing euphoria in the popular press about technological development. "Instantaires," *Time* dubbed the new breed of high-tech entrepreneurs. As the prime source of wealth switches "from the business of making things to the business of thinking up things," echoed a *Newsweek* cover story, the result is "a drastic reduction in how long it takes to get rich." Investors no longer feel the need to see a long string of profits when a company goes public, the story added: "Who needs to show profits? It's the idea that's important."

The Netscape phenomenon has scholars and business entrepreneurs alike trying to grapple with its implications. In the judicious words of William Wresch, computer scientist at the University of Wisconsin, the very fact that Netscape's shareholders would gamble so much capital "tells us we are at a major moment in the commercial development of information."

At times it seems that Adam Smith's hallowed eighteenth-century notion — that the wealth of nations rests on a tripod of capital, labor, and mineral resources — has been displaced by the new triumvirate of patents, trademarks, and copyrights. Some in academia are beginning to take the shift seriously. In 1997 the University of California at Berkeley created an academic position at its business school for a Distinguished Professor of Knowledge. As the *New York Times* reported, the new academic position reflected a national trend in business schools around the country, "groping to enhance their role in an economy thriving less and less on the production of things and more on the production of ideas."

It is not surprising that these trends make many people uneasy. When Bruce Vermazen, chair of Berkeley's philosophy department, heard the news of the new academic position, he quipped, "I liked it better when we made steel. Knowledge seems like kind of a shaky industry."

Of course, Vermazen is using hyperbole to make a point. The world still makes vast quantities of steel. But as Walter Wriston, former chair of Citicorp, notes in his 1992 book, *The Twilight of Sovereignty,* even the manufacture of this archetypal product of the old industrial age now depends as much on information as on the raw materials and labor that go into it. As Wriston puts it, "We build houses and offices and factories from information, we sow, fertilize, and harvest our crops with information and we move our most precious — and some of our most common — possessions on highways of information."

Wriston's statement makes Vermazen's apprehension about the shakiness of the knowledge economy the more noteworthy. Corporations are being advised that their factories, real estate, and materials are less valuable than their intellectual property assets. According to the Tofflers, for instance, "All economic systems sit upon a 'knowledge base.' All business enterprises depend on the preexistence of this socially constructed resource. Unlike capital, labor, and land, it is usually neglected by economists and business executives when calculating the inputs needed for production. Yet this resource is now the most important of all."

For some time, researchers have substantiated the growing importance of knowledge assets in the economy. When economists at the Brookings Institution in Washington surveyed U.S. manufacturers in 1982, they determined that physical assets such as factories, property, and equipment made up 62 percent of companies' total market value, with the rest of the value represented by proprietary knowledge. Ten years later the researchers determined that physical assets accounted for only 38 percent of the companies' value, with the remainder consisting of the firms' intangible knowledge assets.

But how can anyone hope to accurately assess the value of ideas, concepts, and know-how? To appreciate the tenuous nature of an economy based on knowledge assets, think about your own assets. You may own a car, a house, some shares of a mutual fund, but what are your personal intellectual property assets? What kinds of special know-how might you claim? Maybe you know some fly-fishing techniques or have a working knowledge of car engines. Maybe your neighborhood contacts afford you a good command of local politics. Or maybe you inherited your grandmother's secret cookie recipe.

When brought down to the personal scale, the problems posed by these assets become clear immediately. All of these personal bits of intellectual property may have some material worth, but it is very hard to assess their value. Under the right circumstances, some of them could clearly prove quite valuable. Your knowledge of fly fishing could be worth a great deal to overworked executives desperate to escape the city. Your understanding of local politics could be indispensable to someone running for city council. And if your grandmother's name was Mrs. Fields, your cookie knowledge could found a small empire.

But, as all these examples make clear, knowledge has value only when it is put to work. In the late 1800s the British biologist and writer Thomas Huxley recognized this when he wrote that "the great end of life is not knowledge but action." In all these cases, the notion of the specific value of the knowledge is virtually meaningless outside of the context in which it is employed. And even within that context, it is hard to determine its worth as a component of the things it enables us to do. This reality poses a recurrent and vexing dilemma for a knowledge-based economy.

What *Is* Knowledge Worth?

Not surprisingly, given the complexity of knowledge and the difference between knowledge assets and more tangible forms of property, the situation has led to profound confusion. Analysts offer an odd and highly polarized collection of opinions. Esther Dyson, a high-tech analyst whom the *New York Times* has called "the most influential woman in all the computer world," is so taken by the ease with which information can be distributed and copied on the Internet that she reaches the highly dubious conclusion that intellectual property will come to have little or no intrinsic value. Companies should "distribute intellectual property free," Dyson contends, "in order to sell services and relationships."

In Dyson's model of the emerging economy, workers will be paid for services and artists for performances. A work of fiction, for example, will have no value except as an advertisement offering readers the chance to establish greater personal rapport with the author. Dyson likens the scheme to the access to politicians gained by those who attend $500-a-plate dinners and more intimate $10,000-a-head receptions. Among many other problems, the model assumes that people will be more interested in paying for one-on-one access to a novelist than in reading his or her work. It makes little sense in the realm of art or in the business world.

Dyson maintains that intellectual property and all forms of what she calls "content" will lessen in value. But others reach a far different conclusion. Harvard Business School professors Jeffrey Rayport and John Sviokla, for instance, advise companies to exploit what they call "the virtual value chain" by amassing "digital assets." As Rayport and Sviokla note, once firms control assets such as computerized databases of information related to their business, they can "reharvest" them repeatedly through "a potentially infinite number of transactions." In contrast to Dyson, Rayport and Sviokla make intellectual property seem tremendously valuable indeed. And even a cursory look finds an increasing number of high-tech companies heeding their advice.

If analysts stand divided on the basic question of the value of knowledge assets, they are similarly polarized about the way these assets function in the emerging economy. Most economists have argued that knowledge usually functions as an "undervalued externality," a kind of forgotten asset that benefits businesses but doesn't show up in the chart of accounts. But some, like Jim Rebitzer, an economist at Case Western Reserve University, have shown that some firms already do account for knowledge assets, as when a law firm decides which associates to make partner by considering the client lists each associate commands.

Even expert analysis of the broad economic effects of the intellectual property system is murky. Kenneth Arrow, a Nobel laureate in economics, set forth the classical position, asserting that without a strong system of intellectual property rights, too little information will be generated because producers will not be able to capture its true value. But other economists convincingly argue that a weak intellectual property regime can cause *too much* information to be produced by separate groups vying to gain some temporary advantage in trading.

As perplexing as these debates are, the predominant tone of much of the analysis is even more worrisome. Dyson is one of a widely accepted group of observers who strike a triumphalist note, portraying the knowledge-based economy as a largely trouble-free land of opportunity whose emergence is inexorable. "Of course, this new world will distribute its benefits differently than how they are distributed today. But as long as the rules are the same for everyone — and predictable — the game is fair," Dyson claims, with a characteristic laissez-faire undertone. It is hard to know what to make of such a comment. As Dyson should well know, the difficulties rest precisely in choosing what rules to apply, not just in implementing them evenhandedly.

Even more ominously, the Tofflers warn that as we wrestle with the transition to a knowledge economy, "it is the attempt to block such changes, not the changes themselves, that raises the level of risk. It is the blind attempt to defend obsolescence that creates the danger of bloodshed." That argument epitomizes a type of technological determinism that is not just wrongheaded but alarming.

The Tofflers may well be right that the danger of bloodshed exists in the transformation now under way. But they couldn't be more wrong in their identification of the cause. These authors deride any resistance to the juggernaut of the knowledge economy as "blind" and bellicose. In fact, debate about how to apportion the world's knowledge assets is both legitimate and welcome. Only by directly confronting some vexing aspects of the new economy will the public be able to build a system that preserves notions of justice in this new sphere. It won't happen by itself. And the battle lines are just now being drawn.

On one side of the debate, the developed nations, led conspicuously by the United States, seek trade policies that secure dramatically increased private control over intellectual property. These nations affirm the importance of strong, worldwide intellectual property laws in the name of free trade, while at the same time muscling other nations to respect the far-reaching conceptual claims of U.S. companies and research institutions.

The United States is the world's largest producer of intellectual property, including everything from computer software and hardware to pharmaceuticals and movies. Already, by some estimates, more than a quarter of the country's total exports rely on intellectual property rights, and the percentage is sure to grow in the years ahead. Figures compiled by Microsoft indicate what is at stake in software, for instance: some 75 percent of the world's sales of prepackaged software comes from U.S. companies, and the 100 largest U.S. software firms earn more than half of their revenues from offshore sales.

Given this enormous share, it is understandable that U.S. companies have long complained about losing tens of billions of dollars annually to the piracy of their intellectual property. As a result, U.S. trade strategy — in international negotiations from the General Agreement on Tariffs and Trade (GATT) to the 1992 International Convention on Biodiversity — has been to emphasize the importance of protecting that property. "Right now the patent is a sovereign device," explained Lawrence Goffney, who was assistant commissioner at the

U.S. Patent and Trademark Office before returning to the private sector in 1998. "But boundaries are falling down every day. Pretty soon we will have a worldwide patent system." Goffney candidly stressed that the prospective "global reach" will be good for U.S. patent holders.

Employing a more international focus, Richard Barnet and John Cavanagh, in their book *Global Dreams*, estimate that more than 80 percent of the patents held in Third World countries are owned by foreigners, mostly by global corporations; of these, only some 5 percent are actually used in production in those countries. As the system sanctions the ownership of ever more intangible assets, Barnet and Cavanagh charge, "more and more information that used to be freely available is free no more."

Many nations hold that the U.S. stance of ardently protecting the intellectual property of large multinational firms is self-serving and will lead to a regime that needlessly undermines local industries and reinforces disparities and inequalities. What multinational firms and their patent lawyers often construe as protectionist and unfair trade practices can also be seen in a radically different light as legitimate efforts to allow alternative cultures of knowledge to thrive unthreatened by the global market regime. Thailand, for example, has tried to protect its native healers from multinational companies that claim exclusive ownership of their shared medicines and healing techniques. France has also taken an outspoken stance in trying to preserve its film industry from the hegemonic onslaught of Hollywood.

Disputes over intellectual property have already caused an uproar in many nations. In Ecuador, disputes with the United States over intellectual property led the opposition party to occupy the parliament in 1996. Concern over intellectual property provisions during the GATT negotiations spurred enormous demonstrations in India as well. As many as 500,000 farmers in southern India demonstrated to prevent seed companies from privatizing varieties of crops and natural pesticides that have been freely shared for millennia. One faction even ransacked the local offices of a multinational seed company.

The growing dissatisfaction on the part of many international players with the U.S. position was expressed clearly in a 1997 letter to U.S. Secretary of State Madeleine Albright signed by more than eighty representatives of nongovernmental organizations in twenty-one countries. Accusing the U.S. government of using strong-arm trade tactics in the realm of intellectual property issues, the letter complains that too often the United States acts as a "power-broker for commercial interests." As the signatories put it, "By using a 'might makes right' bludgeon, U.S. diplomacy encourages trade wars and destabilizes fragile economies, democracies, and ecologies."

The current global debate over apportioning intellectual property is not surprising. Economic historians have long understood that nations have purposefully employed weak intellectual property laws to jump-start indigenous industries and catch up with technological leaders. In a classic example more than two centuries old, American officials willfully ignored British patents on industrial machines in order to bolster the incipient U.S. patent system.

Today's developing nations are understandably tempted to employ this strategy to protect particular sources of indigenous knowledge or to nurture local technological development. The intellectual property laws favored by the United States make it extremely difficult for these countries to establish their own base in emerging fields such as biotechnology or software development. In such a regime the fledgling local industries have to contend with a daunting array of royalty payments and legal permissions, not to mention the formidable technological hurdles involved.

In one prominent example, disputes over intellectual property have repeatedly brought the United States to the brink of imposing economic sanctions on China. In 1995 imposition of what Mickey Kantor, who was then the U.S. trade representative, characterized as "the largest sanctions in retaliation in American history" was averted just hours before it was scheduled to take effect. Yet for the United States, China represents only the most visible example of strained relations over intellectual property. In the spring of 1997 the Clinton administration threatened trade sanctions against Denmark, Sweden, Ireland, and Ecuador unless these nations increased protection for U.S. intellectual property assets. That same year the United States placed a number of nations — including Argentina, Egypt, Greece, India, Indonesia, Paraguay, Russia, Turkey, and the entire European Union — on probation for similar violations, warning that it would impose future trade sanctions unless these nations did more to protect intellectual property.

From the medicine-rich rain forest of the Amazon to the blossoming software industry in India, international disputes over the value of intangible knowledge resources are only beginning to come to the fore. But make no mistake, they will spawn heightened tension and conflict as poorer nations refuse to be denied rights to their indigenous sources of knowledge.

The cause championed by many people in the world's poorer nations should give us all pause for the same reason: namely, that intellectual property is not truly property at all. Information, knowledge, and techniques exist in a gray area — what some have termed a liminal space — between the marketplace and the polity, between the box office and the soapbox, between the boardroom and the classroom. Any schemes that foster the commercialization of intellectual property at the expense of free sharing of information and ideas will ultimately fail: they will either be met with resistance or they will eventually choke themselves out.

3
Staking Our Claim

GIVEN THE ENORMOUS CHANGES facing the economy and the high global stakes involved, how can we reconstruct the current system to more closely reflect our aspirations for the twenty-first century? As a growing constituency is coming to realize, an emerging economy fueled by knowledge assets requires a system that reflects their intrinsic difference from other kinds of traditional resources.

First, knowledge assets are not quantifiable in the way that land is. If you doubt this, consider the arcane and almost impenetrable way the U.S. Patent Office catalogs claims on knowledge. The classification system shows clearly that intellectual property lines are often nearly impossible to define. It is often even difficult to identify abutters in such virtual terrain.

There is an even more fundamental difference, exemplified clearly by physical assets and manufactured goods such as oil, rice, and running shoes. All of these are finite items in that use depletes them. And, like land, because their availability is limited, traditional goods cannot be easily shared among all who desire them except through a distribution system such as the free market.

Knowledge, by contrast, is not finite in the same sense. It is not depleted by use. On the contrary, books, software programs, or medical procedures lose their value and utility when they are *not* used, when they are sequestered or (perhaps deservedly) neglected. The utility of knowledge assets rests in their exchange and propagation. The always poetic John Barlow captures an aspect of this when he likens the need for an unfettered, healthy flow of information and knowledge to the need of some shark species to keep swimming — for they will die of suffocation if they stop.

People get confused about the limitless nature of knowledge resources because the development of new knowledge often requires a significant initial investment of capital. Software designers and genetic engineers may spend millions of dollars on research and development. Of course, these costly efforts merit recompense. But unlike traditional assets, once a new piece of knowledge exists, it incurs virtually no marginal costs from its ongoing use or dissemination.

Put simply, existing knowledge can usually be shared without making the giver any poorer. Economists who have pondered this feature sometimes call it a case of "nonrival consumption." My consumption of a good, in other words, does not leave you with any less of it. A related quality is encapsulated in the old axiom that if you give someone a fish you feed them for a day, but if you teach them to fish, you feed them for a lifetime. An important aspect of this example is that if I give you a fish, I am one fish poorer. But I am in no comparable way impoverished by showing you my fishing techniques, even if we consider them to be highly valuable. Local fish stocks may eventually wane if too many "educated" fishers depend on the same waters, but even this problem largely evaporates in a knowledge-based system; technological innovation draws from a pond of concepts and ideas that are as unlimited as the human imagination.

Software programmer Richard Stallman illustrates the point. The development of computer software may have cost something, he says, but once it exists, the economic equation changes. He notes, "If more people use a software program, it means that the program contributes more to society. You have a loaf of bread that could be eaten either once or a million times."

This infinite, bountiful, "nonrival" quality of knowledge assets is equally true for software, genetic information, and virtually all formulas, techniques, and languages that might be created, developed, or discovered. This feature forms the cornerstone of the enterprise we call education. As an example, let's say that a doctor develops a lifesaving medical procedure. The more colleagues that doctor teaches the procedure to, the more patients around the world will be cured. Not only do the world's inhabitants benefit tremendously from the dissemination of this knowledge, the original inventor can still perform the procedure effectively even when the entire world's medical practitioners are apprised of it. This single, particular knowledge asset can, in essence, satisfy the needs of an infinite number of users.

The fact that knowledge is not depleted by use, that on the contrary it can flourish and grow only when it is exchanged and shared, should compel us to reconsider some aspects of the way useful knowledge is apportioned in the global economic system. However, just as advances in telecommunications and data transmission are beginning to make possible a new kind of enlightened attitude about sharing knowledge more fully, we find the exact opposite effect: a powerful sector of knowledge moguls, or technological titleholders, is trying to significantly tighten control over these new kinds of assets to create new monopolies.

Rather than resorting to traditional and tired political, legal, and economic frameworks in reviewing various high-tech sectors, I offer three separate but interrelated tools for understanding the changes now under way in the knowledge economy: productivity, equity, and democracy.

PRODUCTIVITY

Our system of intellectual property protection, and the patent system in particular, is widely justified as a tool for spurring innovation. Innovation, in turn, is considered a worthy endeavor in large part because it increases human productivity and thereby generates wealth. Economist Lester Thurow made the classical argument in a 1996 essay on the piracy of intellectual property. Patents and copyrights, Thurow says, involve an inherent tension between providing an incentive for new discoveries and spreading the discoveries around. As Thurow explains, "The faster knowledge is spread, the faster human welfare will rise. If spreading knowledge is the sole goal, the patent and copyright system should be abolished." But if the point of the system is to help spur innovation, he says, "we need strong intellectual property laws."

Such a trade-off challenges our democratic rhetoric favoring public access and opportunity. But it remains highly debatable, especially in certain high-tech sectors, whether the present system does spur innovation. Few observers can see anything positive, for instance, in the growing tangle of intellectual property litigation, especially in fields like software development. Such disputes ultimately waste resources, lower productivity, and often stifle new development.

Observers have recognized the pitfalls of the productivity argument for some time. The brilliant mathematician and father of cybernetic theory, Norbert Wiener, wrote about the problem back in 1954, arguing that "the value of a piece of scientific work only appears to the full with its further application by many minds and with its free communication to other minds." As Wiener put it, "Here any secrecy or any rights of property possession will naturally have the effect of making people shy off from a preempted field of work." It seems we have forgotten this basic lesson.

Because the productivity argument is a bottom-line rationale for the system, the chapters that follow critically ask: is the existing system spurring innovation? To the extent that the system actually stifles innovation, as it does in many cases, it becomes particularly hard to justify, even on its own terms.

EQUITY

Issues of equity cut in many contradictory ways through the thicket of intellectual property protection. Americans by and large revere market competition as a means to boost the supply of goods and bring down prices. But even most diehard free market advocates draw the line at making use of a competitor's innovation. We teach our children that it is unfair to copy someone else's work and then pass it off as one's own. In this social construct, the appropriation of new ideas goes beyond competition; we call it plagiarism or piracy.

On a broader scale, economists call the phenomenon the free-rider problem. Free riders are seen to compete unfairly by profiting from the fruits of a development to which they did not contribute. If you have spent millions of dollars to develop a product, for instance, you will understandably be distressed when a rival produces a comparatively inexpensive knockoff of your hard-earned design.

But the line between "idea snatching" and sanctioned idea sharing is often fuzzy at best. As many commentators have noted, the system must successfully balance individual and group interests. Increasingly, people around the world are raising difficult questions about where innovations derive from in the first place and whether our system adequately acknowledges and compensates those in the milieu from which they arise. Does the pop singer owe something to the indigenous musician whose melodies she expropriated and profited from? Does the screenwriter owe a debt to the subject upon whom he based the script? Is a software designer within her rights to build upon an existing program and patent the new program as her own? The difficulty of answering such questions is evident in the seemingly capricious court verdicts in intellectual property cases and the problems of enforcing antitrust laws, as we pick and choose whether and when we deem various types of competition to be fair.

The emergence of a knowledge-based economy forces us to realize that we are all to some extent free riders. On the one hand, the present system insists, often despite counterfactual evidence, that important breakthroughs occur through the vision of a single individual who creates something new from scratch. This view explains why many kinds of scientific advances and artistic works are credited and handsomely rewarded. Yet as historian Gar Alperovitz notes, "What we accomplish stands atop a Gibraltar of technological inheritance. Seemingly contemporary transformations inevitably build on knowledge accumulated over generations."

Such a view forms the basis for a strong public-interest claim to some of the fruits of technological know-how. In the United States, taxpayers contribute some $60 billion annually to underwrite the bulk of the nation's basic research and development at universities, research institutes, and national laboratories. To a degree that we are only now coming to appreciate, we all own a piece of the technological rock. And yet we allow private firms and individuals to claim more and more of the proceeds from our public investment.

DEMOCRACY

Democratic arrangements are distinguished by the access and accountability they afford their participants. I can use my local municipal park freely, and it cannot be paved over to make a private parking garage unless, after a public hearing, such a plan is agreed to by a local majority. Some elected representatives may secretly want to turn the park into a parking garage, but these officials are, at least technically, accountable to me. I have a right to know about their scheme and to have a say in the matter. Yet I have none of these rights over the empty, privately owned lot adjacent to the park.

Despite the power of democratic arrangements, we live in a time of privatization unprecedented in modern history. In some cases private management of public problems works effectively. But, even granting the highly debatable notion that turning the management of everything from roads to prisons over to private companies is an efficient solution, privatization can dramatically change the quality of shared institutions.

The broad trend to apply market solutions to political problems puts pressure on the administrators of public institutions from libraries to high schools.

It also has a dramatic impact on less tangible realms, such as the Internet or the emerging map of the human genome. As we survey examples of privatization in the intellectual property arena, the key question is this: by allowing private ownership of formerly public assets, to what extent do we threaten access and public accountability?

Corollaries to the privatization of intellectual property assets can be seen all around us. Take, for instance, the licensing arrangements by formerly noncommercial groups that have now become commonplace. The summer Olympics of 1996 in Atlanta was widely dubbed the Coca-Cola Olympics, as the company managed to take private sponsorship of this international event to a new level, putting its logo on every conceivable venue. And New York City has hired International Management Group (IMG), a Cleveland-based agency that represents superstar sports clients like Tiger Woods and Evander Holyfield, to sell corporate sponsors exclusive rights to advertise and market their products in the city's municipal spaces. The plan is to sell space for corporate logos on everything from trash cans to city-owned vehicles. IMG Senior Executive Vice President Robert D. Kain jokes about whether the city will sell Pepsi or Coca-Cola the "exclusive pouring rights" for Central Park. But in today's climate the idea is not particularly farfetched. Scores of public schools around the country have sold precisely these rights. In the unending quest to build market share in the next generation, Coca-Cola and PepsiCo have paid schools between $200,000 and $1 million each, plus a percentage of vending machine sales, for the right to be the exclusive soft-drink vendors on school property.

Such private licensing arrangements may seem relatively innocuous, but, to the extent that they erode a commonly held asset, they can be significant indeed. Roughly three decades ago, in an article in *Science*, biologist Garrett Hardin identified what he called "the tragedy of the commons." He documented that without external safeguards to protect them, shared public resources, like common grazing lands, will eventually — and inevitably — suffer from private overuse. While Adam Smith identified an "invisible hand" that leads private self-interest unwittingly to serve the common good through the efficient exchange of goods in the marketplace, economist Herman Daly points out that this mechanism backfires where shared public goods are involved. In these cases, what Daly terms "the invisible foot" leads "private self-interest to kick the common good to pieces."

Using the lens of democracy, the next chapters will ask: To what extent is the invisible foot at work in high-tech realms? Is a commonly held asset eroded, for example, when a doctor claims private ownership of a widely used medical procedure and demands royalties from his colleagues? Or when a university allows private sponsors to dictate the topics for its faculty and graduate students to study? Or when a national park sells a private firm exclusive rights to the genetic information contained in the park's plants or microorganisms? If so, are there clear democratic mechanisms to halt the process if a majority so desires? Unfortunately, we often learn that it is relatively easy to erode commonly held assets but more difficult to build them back.

The issues arise in every facet of our participatory democratic system. "The press and the public are being slowly blinded," says Bill Kovach, respected journalist and curator of the Nieman Foundation at Harvard, about the privatization of formerly public institutions. As he notes, a publisher in Mississippi has claimed exclusive rights to distribute and sell the electronic version of the state's laws. In Newton County, Texas, journalists have been denied access to prisoners in the privatized jails; the managers have claimed their prerogative to establish a policy of "no media contact with prisoners." The public's access and voice are clearly limited by such private arrangements. So is the system's accountability: freedom of information laws do not cover private businesses.

It doesn't take a rocket scientist — or in this case a political scientist — to note that we are witnessing a collision between two venerable traditions: capitalism and democracy. Hints of the collision are swirling in every field.

International financier George Soros identified a piece of this picture in a thoughtful critique entitled "The Capitalist Threat," which appeared in the *Atlantic Monthly.* "Insofar as there is a dominant belief in our society today, it is a belief in the magic of the marketplace," Soros writes. "The doctrine of laissez-faire capitalism holds that the common good is best served by the uninhibited pursuit of self-interest." Yet this quintessential capitalist, an entrepreneur who has made a multibillion-dollar fortune in the international currency market, has come to realize that the doctrine is false. As Soros puts it, "Unless it is tempered by the recognition of a common interest that ought to take precedent over particular interest, our present system — which, however imperfect, qualifies as an open society — is liable to break down."

It is, after all, our democratic instincts, emphasizing public access and opportunity, that have brought into being many of the qualities of society that we prize. If not for the democratic tradition, dating back to the Enlightenment, would national parks exist? Would *any* public parks exist? Would we have public libraries, museums, or public education? Probably not. Try as we may, we simply cannot justify these civic institutions on the basis of market efficiency alone. As James Fallows has written, "Does anyone believe that Central Park would exist if the market had been left to determine the most efficient use of land in Manhattan?"

As we look for guidance, it is fascinating to see how fully aware the nation's founders were of this potential collision as they established our democratic system. On the one hand, they included the makings of a strong intellectual property regime in the very first article of the Constitution. Article I, Section 8, provides that "the Congress shall have Power . . . to promote the progress of science and the useful arts, by securing for limited times to authors and inventors the exclusive right to their respective writings and discoveries."

Yet they also expressed grave reservations about the patent system they were sanctioning. Thomas Jefferson, the nation's first patent administrator, initially resisted the idea of the patent system itself, saying he found the concept of even limited monopolies repugnant. And although he later warmed to the system, he never sought to patent any of his own inventions. Neither did Benjamin Franklin or his well-known successor in electrical science, Joseph Henry, the Smithsonian Institution's original secretary. Henry said no "true man of science" would resort to patenting discoveries. He regarded patents as corrupting and antithetical to a spirit of free inquiry.

Jefferson, Franklin, and the other Enlightenment designers of American democracy came to support patent and copyright laws largely as a means to assure the widespread distribution of ideas into the libraries and minds of their young republic. The sentiment is captured in George Washington's farewell address, on September 19, 1796, when he urged the young nation to "promote then, as an object of primary importance, institutions for the general diffusion of knowledge."

James Madison beautifully affirmed that position in 1822, writing that "a popular government, without popular information or the means of acquiring it, is but a prologue to a farce or a tragedy: or perhaps both." As he stated, "Knowledge will ever govern ignorance; and people who mean to be their own governors must arm themselves with the power which knowledge gives." The words resonate with tremendous vitality at this current historic moment. It is helpful to keep Madison's wise counsel in mind as we consider the disputes in medicine, software and drug development, agriculture, and the information industry in the chapters that follow.

As I will ultimately argue, even a nation that champions private property can venerate a national park system that preserves some land for shared use, land sanctioned as too important to the collective heritage to be offered for individual sale. Zoning and antitrust laws — both direct curbs on unfettered private ownership — have also been essential and effective tools for fostering a more healthy and equitable system of private enterprise. But no analogous mechanisms yet exist on the unbridled knowledge frontier.

Taken together, the dispatches offered from a variety of high-tech frontiers in the next several chapters lay the groundwork for the argument that we urgently need an enlivened debate about how to preserve our democratic processes from inroads made by the dramatic expansion of ownership claims. Our goal should be to protect vital knowledge resources whenever possible so they can best be shared to benefit all of us rather than to enrich a select few.

Part II

Dispatches from the Knowledge Frontier

4

The New Medical Licenses

JUNE 1993, RANDOLPH, VERMONT. Until a lawsuit threatened to upset his career, Dr. Jack Singer's professional life was a picture of composure, precision, and order. Singer's medical practice and his teaching career at Dartmouth Medical School were thriving. The academic year had just ended, and spring had finally worked its way north to Vermont. But one day, as Singer flipped through the mail, he noticed an envelope from an unfamiliar law firm.

Doctors universally fear malpractice lawsuits and pay sky-high insurance premiums to guard against them. An eye surgeon at a satellite of the Dartmouth-affiliated Lahey Hitchcock Clinic, Singer was always cognizant of the perils involved in operating on people to restore their sight or correct their vision. Meticulous and steady, he was well suited to his work. His reputation in the field was growing, and he knew of no outstanding complaints against him. But the possibility of an unexpected malpractice claim always lurks as a hazard for any physician. He quickly opened the letter.

As it turned out, Singer was faced with a lawsuit, but not for malpractice. This lawsuit was threatened precisely because of the success of an operation Singer had recently perfected to remove his patients' cataracts, using a specially shaped incision that requires no stitches to heal. According to the letter, a surgeon in Arizona had recently patented the cataract operation and was exercising his legal claim to exclusive rights over it. Formal, legalistic language notified Singer that if he wanted to use the no-stitch cataract operation on his patients he would have to agree to a licensing arrangement, paying royalties to the Arizona surgeon each time he performed the procedure.

The letter notified Singer that he could expect to pay royalties amounting to somewhere between $2,500 and $10,000 annually, depending on how many cataract operations he performed. Even worse than the imposition of royalties, though, was the letter's warning that if Singer continued to perform the operation without obtaining a license, he would be sued for patent infringement. He would be forced to deny his patients the latest advance in cataract surgery — a technique Singer had already mastered and had begun to teach to fellow ophthalmologists.

Singer was stunned and shocked as he skimmed the letter, but his shock quickly turned to anger, shattering his normally unflappable demeanor. "Before I knew it," he recalls, "I had nearly slammed my fist through the counter in my office." What was happening to his profession when practitioners were claiming to own medical procedures, he wondered incredulously. Could this seeming extortion possibly be legal? And why should he have to wade into a legal quagmire when his career was going so well? Singer's deft and innovative cataract work had led to invitations to lecture and teach at professional gatherings around the world.

Singer had seen a lot of changes in his eight years as a physician. Managed care, HMOs, and for-profit hospital corporations had begun to dominate the health profession. In ophthalmology an increasing number of legal disputes had arisen between practitioners and the manufacturers of high-tech lasers and other equipment upon which ophthalmologists relied. But the prospect of doctors charging each other royalties for doing their job as healers went far beyond any broad concerns he might have had about the direction of health care policy in the 1990s. It was an affront. The very idea of "licensing" a medical procedure ran counter to everything Singer believed about his profession. After having spent the better part of a decade in medical training, he felt that his state license to practice was the only medical license he was interested in.

As Singer would testify nearly three years later before Congress, he was not one to worry idly about the future of the medical profession and warn that the sky was falling. There was nothing hypothetical about his experience that June day: "The sky actually did fall on me."

In an initial state of denial, Singer thought the matter might be a mistake, a misunderstanding between two medical colleagues that could be cleared up with a personal response. He soon learned otherwise. The case would eventually cost Singer, the Lahey Hitchcock Clinic, and the supporters who contributed to Singer's legal defense fund more than half a million dollars in legal fees. It would inspire an outpouring of concern from sixteen separate organizations of medical professionals and eventually even spur federal legislation.

The implications of Singer's situation troubled many disparate parties. If doctors started suing each other for royalties over new medical procedures, patients would have to either absorb the additional costs of royalties or, worse, be denied the latest advances in treatment by health care providers who were not licensed to offer them. Most everyone recognized a problem brewing, especially medical practitioners and their public advocates, who envisioned a horrendous tangle of litigation that would pit colleagues against one another and draw further ire from a public already critical of the excesses and spiraling costs of medical care.

The emergence of cases like Singer's caused even some patent lawyers to worry aloud — a group that tends to remain sanguine about the robust power of the patent system to work out its own problems over time. William D. Noonan, for example, a patent lawyer based in Portland, Oregon, who also holds a degree in medicine, said that despite his faith in the merits of the patent system, some things deserve special protection. "Suppose a person is choking on food in a restaurant and needs to have his airway cleared by the Heimlich maneuver," Noonan suggested. "If the Heimlich maneuver had been patented, would one hesitate to save the hapless diner for fear of infringement?" The prospect of such life-threatening hesitation, even more than an increase in actual infringement disputes, Noonan and others suggest, ought to be argument enough to reevaluate the issuance of patents on medical procedures.

Not only would patients suffer from the privatization and licensing of medical procedures, the ethics of the profession faced a significant threat. As Singer's lawyer, Robert Portman, notes, given "the thousands of different surgical procedures that are performed every day, the potential for the proliferation of patents on medical and surgical procedures becomes a frightening prospect." Singer's case, Portman contends, is just one indication of "the potential havoc that medical-procedure patent holders can wreak on the delivery of medical services."

Singer's case drew the attention of the American Medical Association (AMA). In 1995 the AMA's then-president, Robert McAfee, emphasized the threat the case posed to the patient-centered altruism of the Hippocratic oath and its call to teach the craft of medicine "without fee or covenant." As he said, "It's part of the Hippocratic Oath to freely share information that you have with your colleagues." According to McAfee, "The history, excellence, and tradition of medicine has been that whenever a new procedure occurs and is proven effective, it is imperative to share that knowledge with the world at the earliest moment." The oath that bears the name of Hippocrates — the Greek "father of medicine," who practiced hundreds of years before the birth of Christ — has been taken by physicians entering the profession for more than a thousand years, including its provision to share new insights freely among practitioners old and new. Unfortunately for Jack Singer, however, the Hippocratic oath, still overwhelmingly endorsed by doctors around the world, does not carry the force of law.

"That's an Invention?"

If Singer's case hit a nerve, it was largely because many in the medical field recognized that he was not simply an unlucky anomaly, a practitioner targeted by a lone, vindictive colleague. Rather, the case augured a startling new trend. In 1993 the Patent Office was already awarding scores of patents each month on medical procedures, and by early 1996 the number had reached an unprecedented one hundred per month. Medical specialists of all kinds, from radiologists to urologists, began to receive threats from colleagues similar to the one that changed Jack Singer's life. "Proliferating medical-procedure patents," the conservative trade journal *Medical Economics* complained in 1996, "entangle not only surgeons, but a wide range of practitioners." And because patent examiners, who are seldom medical practitioners themselves, base their decisions on searches of published work that often poorly reflect the unfolding state of medical knowledge, patents are frequently granted for procedures that are not particularly novel or even noteworthy. As *Medical Economics* pointed out, many of the ownership claims cover skills most doctors are expected to learn during their medical residencies.

One doctor, for instance, owns the rights to a basic technique for suturing the stomach to the intestines. Another claims exclusive rights to the common technique of making slits in a skin graft to expand it. Individual practitioners already own such simple procedures as treating a nosebleed with a catheter wrapped in gauze. One doctor even patented the common practice of applying the anesthetic lidocaine to the skin to treat nerve pain associated with the herpes skin infection known as shingles.

In one example that occurred around the time of Singer's legal battle, radiologists across the country received letters attempting to exact royalties on a patent covering a technique for determining the sex of a fetus at age twelve to fourteen weeks with ultrasound. The procedure's patent — legitimate and still technically valid — boils down to visually distinguishing male genitalia from female. Not surprisingly, the American College of Radiology has condemned the patent. Many in the field have also publicly scoffed at it. As Chris Merritt, a radiologist at the Ochsner Clinic in New Orleans, says, "It's like saying you have a secret method for distinguishing the gender of patients when they take their clothes off for a physical. That's an invention?" Another likened it to a patent on telling your right hand from your left.

The matter can't simply be laughed away, however. As unbelievable as it sounds, John D. Stephens of San Jose, California, a radiologist who specializes in prenatal diagnosis, does in fact hold a twenty-year monopoly over the technique. Stephens is fazed by neither the seemingly obvious nature of this "invention" nor the fact that it relies completely on the work of untold thousands of medical engineers who developed and refined the ultrasound imaging devices needed to perform the procedure. He says the patent is merely his rightful means to protect his intellectual property. As for the Hippocratic ethic, Stephens calls it "naive and out of date."

As it turns out, the basis for his patent claim itself is questionable. After his demands for royalties sent colleagues scrambling for legal advice, Stephens backed off from the threat. But it is not an incidental or trivial matter that our system officially sanctions an individual's ownership rights over such a process. Even if they are not enforced, personal ownership claims to formerly shared techniques can have serious consequences.

For one thing, the difficulty of establishing the origin of a widely shared medical procedure makes the system ripe for abuse by a new kind of pseudo-inventor: someone who skims claims from the vast pool of freely exchanged medical knowledge. As Singer puts it, "Everybody can start claiming ownership of slight variations of methods available in the public domain," resulting in a "profound chilling effect" on medical practice.

"Once patenting enters the picture," Singer explains, "physicians inevitably realize that they face a potential legal risk every time they introduce a new procedure or a modification of an existing procedure into their practice." The result runs precisely counter to the patent system's intended goal of fostering innovation. The mere existence of medical procedure patents, Singer says, cannot help but lead physicians to become more guarded about sharing their experiences. They will inevitably fear identifying themselves as targets for patent infringement suits or exposing themselves to charges that they have induced others to infringe the patented procedure. "I know this phenomenon well," he says, "because this is precisely what happened to me." It was only after Singer wrote an article in the *Journal of Cataract and Refractive Surgery* updating fellow specialists about the sutureless cataract procedure that he was threatened with a patent infringement lawsuit.

The story of Jack Singer's troubles actually begins across the continent, well before the publication of his article. On a spring day in 1990, Samuel Pallin, an ophthalmologist in Sun City, Arizona, accidentally discovered a new twist on the cataract operation he routinely performed. As Pallin tells the story, he had made "an upside-down chevron-shaped incision" in a patient's eyeball but didn't have time to stitch the cut because the elderly patient experienced heart problems and had to be rushed to the hospital. Two weeks later the patient's scar had healed without a suture and displayed less scar tissue than usual.

"I had been in this business for years. I had never heard of anything like what I had done," Pallin told the *Phoenix Gazette.* But when he submitted an article about it to the *Journal of Cataract and Refractive Surgery,* the rejection letter called his procedure "yesterday's news." Many other eye surgeons were already using similar techniques, the journal's editors said.

Stung, Pallin decided to try to patent the surgical procedure of making a self-sealing incision in the eye and found he was the first to claim ownership of it. Many eye surgeons, as the journal's reviewers claimed, had undoubtedly begun to employ similar techniques at the time of Pallin's claim in 1990, but it had not been widely enough reported to persuade the Patent Office that its "obviousness to practitioners in the field" — a key hurdle in the quest for a patent — should exempt it from a monopoly ownership claim.

Until 1954 medical procedure patents were entirely and explicitly prohibited. Patent directives and patent-law precedent drew a hard line between devices, like catheters and X-ray machines, which were deemed eminently patentable, and procedures, such as blood transfusions or coronary pulmonary resuscitation (CPR), which were not. The rationale for the distinction was plain: the development of medical machines and instruments often required a significant capital outlay that could justify the allotment of a government-sanctioned monopoly as an incentive that allowed inventors to recoup their costs; the refinement of new treatments rarely entailed such development costs.

Undoubtedly, the Patent Office was also wary of the costs that would attend a proliferation of procedure patents. The costs of new medical devices like X-ray machines and heart monitors have long been identified as a major cause of rising medical costs. Licensing fees for medical procedures would only add to these already substantial increases.

Finally, there was a practical matter: while it was relatively easy to track the number of medical devices manufactured and sold, enforcing a medical procedure patent would be far more difficult and, to some degree at least, was protected by doctor-patient privacy laws. Historically it was difficult for doctors to monitor the medical procedures used by their colleagues. "What could a doctor do, stop people on the street and ask to see their scars?" asks George Annas, a specialist in health law at Boston University. But with the advent of HMOs, vast computer databases of medical records, and a shift toward running medical practices more like businesses, the prospect of tracking infringement for medical procedure patents has become considerably more feasible. With the widespread availability of so much computerized medical data, as an AMA official explained, "it is not surprising that this patenting concept would surface."

Whatever the reason, the Patent Office's distinction between devices and procedures began to erode, in concert with a similar expansion in many other fields of the accepted notion of what constitutes "intellectual property." Beginning with a fateful 1954 patent on a technique to treat hemorrhoids, the Patent Office became steadily more indifferent to the distinction between a device and a procedure. By the early 1990s many doctors had come to realize the potential benefit of laying such claims.

Philosophically, Patent Office administrators have essentially come to side with the view of Dr. Pallin, who defends his lawsuit against Singer by claiming that patenting a medical technique is no different from patenting a new drug or surgical instrument. "We don't think of it as greedy when a scientist gets a royalty for coming up with a new [drug] compound," Pallin contends. "It is ridiculous to say that this is any different." After all, he notes, at least in the United States, "medicine is a capitalist endeavor." However expedient, such a view has ultimately provoked deep-seated divisions among practitioners and policymakers alike.

Shortly after filing for a U.S. patent on the sutureless cataract procedure, despite his initial rejection by the *Journal of Cataract and Refractive Surgery*, Pallin was invited by that journal, along with a number of other doctors and medical researchers, to contribute to a special issue on sutureless incisions in 1991. The issue, which covered both clinical progress and the biomechanics of the cornea's ability to seal itself, included an article by Pallin on his so-called chevron incision as well as a paper by Singer detailing a self-sealing "frown incision" that he had been using and teaching to colleagues. It was Singer's article that alerted Pallin to the fact that others in the field were putting their personal stamp on an operation he continued to believe he alone had invented.

Pallin's patent was awarded in 1992. In July 1993, the month after the letter to Singer, Pallin brought formal legal action against the Vermont doctor, whose article he had read alongside his own.

Singer felt he had little choice but to fight. "It would have been much easier to purchase Pallin's license and let others worry about the problem. But from the beginning, I knew I had to fight this as a matter of principle."

Singer says he guessed Pallin had calculated that a small ophthalmology practice in rural Vermont would probably lack the resources for a costly legal battle, setting a good precedent as Pallin targeted others with his licensing demands. When Singer refused to pay royalties, Pallin sued him and Hitchcock Associates (Singer's branch of the Lahey Hitchcock Clinic) in U.S. District Court in Vermont, charging that they were infringing on his patent and, through publications and teaching, inducing other surgeons to do so as well.

When the matter came to trial in 1995, the legal case hinged on whether Singer could prove that he had been conducting the cataract operation before Pallin filed for the patent. As it turned out, Singer did have documented proof of what lawyers call "prior art." He could show that he had been performing the operation for one slim month prior to Pallin's patent claim, enough to secure a legal victory. By this time, though, Singer's case had caught the eye of Greg Ganske, a plastic surgeon by training who was also a Republican member of Congress from Iowa.

Ganske couldn't believe the situation when he got wind of it. The prospect of medical practitioners suing each other over their procedures seemed so obviously to contradict the established service orientation of health care that he felt Congress simply had to step in. "Medical advances need to be shared for the betterment of all, not the enrichment of a few," Ganske wrote to his colleagues in the House of Representatives.

Ganske's response to the problem was straightforward: he proposed legislation that would simply outlaw patents on medical procedures. As he pointed out, some eighty countries around the world, including Great Britain, Germany, and Japan, already exempt medical procedures from patent protection in order to safeguard the dissemination of new techniques and knowledge.

Ganske's legislation had widespread support within the U.S. medical profession as doctors fearfully watched the unfolding *Pallin v. Singer* case. A 1995 AMA position paper argued that the organization had "considered and rejected the argument that patenting medical procedures may ultimately benefit patient care by providing an incentive for innovators and private investors to support research and development. There is no empirical evidence to support the claim." The report concluded that "the perils of patenting outweigh the benefits" in this arena.

Such unequivocal pronouncements from the powerful AMA and other medical lobbying groups helped garner support for the legislation. But Ganske's plan met a roadblock. The problem, as one legal reporter put it, was that Ganske's proposed legislation "struck terror into the heart of the medical and pharmaceutical community, and particularly the biotechnology community." Groups testifying against the measure included the American Intellectual Property Law Association, the Intellectual Property Law section of the American Bar Association, and the Biotechnology Industry Organization. Besides making the familiar arguments that patents foster progress, biotechnology industry representatives warned that Ganske's bill might harm pharmaceutical companies that use medical procedure patents when they find new uses for an existing prescription drug.

The lawyers also argued that the Ganske bill would invite other countries to weaken their own patent laws at a time when the United States was pushing to make them tougher. Along these lines, opponents of the Ganske bill argued that it would embarrass the United States internationally, possibly even putting the nation out of compliance with the very intellectual property treaty arrangements it was working so hard to force down many nations' throats in the GATT negotiations. An article in the *New York Law Journal*, for instance, argued this position clearly. Citing American intellectual property as "one of the country's greatest assets and exports," patent lawyer Jeffrey Lewis warned of a "domino effect" in which some "countries that have only just begun to enforce patents (or are still being pressured to do so) will follow the example of the United States and systematically exclude certain types of patents on 'policy' grounds."

Ultimately, a curious legalistic compromise won the day, offered by another doctor, William Frist, a thoracic surgeon and Republican senator from Tennessee. The final legislation, signed by President Clinton in the fall of 1996, specifies that doctors can apply for and receive patents on medical procedures, but medical practitioners are exempted from liability if they conduct these procedures in the course of their practice. In other words, patents on medical procedures are acceptable, but lawsuits like the one brought against Jack Singer would henceforth be forbidden.

Numerous medical specialty groups rallied behind Frist's milder proposal, but even this change worries many legal and biotech lobbyists. Some believe that giving doctors immunity from infringement suits will encourage other groups to ask that they too be placed "above the law." However self-serving the view may be, it raises an important and legitimate point by highlighting the problems of inconsistency and blurred distinctions in rules governing intellectual property.

Seen in one light, the advent of federal legislation on the issue of medical procedure patents shows that gross excesses in the control and ownership of knowledge assets can be corrected. On the other hand, though, the congressional fix fails to address the deeper underlying issues: Is the delivery of health care a public service or a business? What aspects of the enterprise should be privately owned and what, for the greater good, must be held in common? How can we reward individual innovation in medicine while protecting the openness and free exchange of information vital to its perpetuation and advancement? As bitter intellectual property fights spread throughout the biomedical field, Congress's actions may come to be seen as a feeble finger in the hole of a weakening dike.

Congress may have forestalled escalating legal battles between medical practitioners, but the compromise bill does not apply retroactively, so thousands of actionable medical procedure patents are still on the books. Even more important, the law does little to address similar problems arising in biomedical research, a far more lucrative branch of the medical field than Jack Singer inhabits, and one in which the most important players are more likely to be large corporations than individual practitioners.

Costly legal disputes over intellectual property are, of course, far more common among biomedical research firms than between individual medical practitioners. And as many people are coming to realize, jockeying over ownership rights currently threatens the same kind of chilling effects and other deleterious consequences that were so clearly brought to light in *Pallin v. Singer.*

Nonetheless, in the effort to spur innovation and protect individual accomplishments, patent law holds research and development to far different standards from those of the daily practice of medicine, even if access to new drugs and therapies with lifesaving potential is curtailed. In fact, several months before Clinton signed the bill to limit patent infringement cases on medical procedures, he signed into law a bill expanding patent rights in biomedicine. "I am pleased," Clinton said at the signing ceremony in November 1995, to sign a bill offering "enhanced protection of biotechnology process patents." The law, Clinton said, will "provide the protection American biotechnology companies need to continue developing new products. American consumers will benefit from improvements in the diagnosis, cure, or treatment of disease."

If Clinton's actions seem contradictory, they exemplify the widespread confusion in questions of intellectual property rights. To see how these issues are affecting the field of biomedical research and to understand the profound inadequacy of the limited congressional compromise, it is worth examining one of the great medical advances of recent years, the development of ex vivo human gene therapy, a medical procedure for which Clinton and Congress specifically sought added private protection.

The Ex Vivo Files

The story came to public attention most dramatically in September 1990, when television newscasters and newsweekly reporters clamored to get the story of four-year-old Ashanthi DeSilva, the world's first patient to receive human gene therapy. W. French Anderson, who led the medical team that performed DeSilva's surgery, claims that human gene therapy represents one of the "great leaps" in the history of human health care, along with such formidable advances as the development of anesthesia and vaccinations. It is still early to know if such a grand assessment will be borne out, but Anderson's credentials lend his views a good deal of weight. Now head of Gene Therapy Laboratories at the University of Southern California School of Medicine, Anderson spent twenty-seven years at the National Institutes of Health, where he earned the unofficial title of "the father of gene therapy." The title is actually official in one important sense: Anderson holds breathtakingly broad ownership rights over the technology.

As the public learned from the barrage of media attention Anderson received during his team's pathbreaking therapy, DeSilva, along with another child, suffered from an exceedingly rare type of genetic disorder: their bodies failed to produce an enzyme called adenosine deaminase, or ADA, which the body needs to produce disease-fighting white blood cells. Because of their severe immune deficiency, an outing to the park meant risking a cold or other viral infection that could likely kill them. The girls were, in effect, quarantined "bubble children." Although 1 out of every 150,000 children born is afflicted with some kind of genetic disorder, the particular type these girls had was known to exist in only thirty other people worldwide. Because the disorder derives from a mutation, or misspelling, in a single human gene, it was a particularly good candidate for the highly experimental therapy.

To treat the girls, Anderson's all-star medical team took certain white blood cells called T cells from the blood of DeSilva and Cynthia Cutshell, who was then nine years old. The team cultured the cells until they multiplied many thousandfold, then mixed the cells with a virus that had been altered to include the normal ADA gene. The virus had been rendered harmless, but it still infiltrated the defective T cells, transferring the corrected ADA gene. When the team injected the girls' corrected T cells back into their bloodstream, the virus continued its work of spreading the normal ADA gene.

The technique Anderson's team used is called ex vivo gene therapy because the cells of the patient are removed and altered in the lab, outside of the body. Other research, known as in vivo gene therapy, which is still in its infancy, uses a virus or other vector to insert genes directly into a patient, so their corrective work can be accomplished entirely within the body.

By all accounts, the ex vivo gene therapy operation was a success. DeSilva now attends school and, after receiving additional altered T cells for more than three years after the initial operation, her body continues to have incorporated the ADA gene into her immune system. Both girls are now living near-normal lives, despite being born with a usually lethal immune disorder, a prospect that led team member Michael Blaese to exclaim at their follow-up evaluations in 1995: "We're absolutely thrilled. We couldn't have asked for a better outcome."

In marked contrast to the attention received by the success of the procedure, however, was the relative dearth of public information about a set of equally portentous developments. Relatively unnoticed except in the scientific and trade press was the filing by Anderson, along with Blaese and team member Steven Rosenberg, for a U.S. patent on every kind of ex vivo gene therapy in humans. The patent covers all ex vivo manipulations in which malfunctioning human cells are genetically altered to produce therapeutic levels of protein outside the body and then replaced — a technique with potential applications ranging from treatments for brain tumors to arthritis. Although the patent does not cover in vivo manipulations and is restricted to the United States, it is still broad enough to have generated widespread comment and controversy within the scientific community.

There is little question that the ex vivo patent covers a medical procedure. It would be more accurate, in fact, to say that it covers a whole new, fertile field of possible medical procedures. In fact, if the patent had been issued after 1996, it might even be construed as falling under the purview of the recent legislation on medical procedure patents. But in deference to the biotechnology lobby, Congress specifically exempted genetic engineering processes even when they move, as the ex vivo patent does, fully into the clinical arena.

Despite a research environment inured to the encroachment of private ownership over conceptual tools and intangible assets, the breadth of the ex vivo patent drew cries of disapproval. "Deep disbelief, I'd say that's what most people feel about the breadth of the patent," Joseph Glorioso, head of the department of molecular genetics and biochemistry at the University of Pittsburgh, told the journal *Nature* when he heard of the patent. "This is analogous to giving someone a patent for heart transplants."

Some observers were willing to dismiss the case as another unfortunate episode in which the Patent Office simply granted an overly broad ownership claim. After all, development costs had to be recouped and the innovative therapy was well deserving of recognition. Furthermore, Anderson stresses his team's functional leap from theory to practice. "What we did was to provide proof in principle that gene therapy can work in humans," he says.

Many others, however, offer a more fundamental critique. Dusty Miller, who played a seminal role in developing the treatment as a collaborator on Anderson's team, calls the breadth of the human gene therapy patent "beyond comprehension." According to Miller, "There shouldn't even be a patent on this, or at the very least it should be far more narrowly defined." From his standpoint, the ex vivo patent amounts to "another big step toward the bizarre world where people stake claims to the natural processes of the human body."

The patent's breadth is only one of a dense thicket of difficulties the complex case raises. Equally vexing — and still a point of contention in this case — is the question of attribution. When the development of a new technique builds upon many years of work by multiple researchers, as it so often does, who deserves the ownership rights?

"The Patent Office has some sort of notion that French and Mike and Steve were sitting around in a room in 1988 and they thought up the idea of human gene therapy," Miller says. "But, of course, this is not at all the case. People in the field had been actively discussing human gene therapy for ten years." As it happens, Miller has good reason to review this point, because his own role is pertinent and problematic.

In the late 1980s Miller and another scientist, Kenneth Culver, helped Anderson and others at NIH develop the ADA gene therapy treatment. Even at the time, Miller was world renowned for his work modifying viruses. By all accounts it was Miller's viral delivery technique that ultimately allowed the medical team to infiltrate DeSilva's T cells with the corrective genes. As *Chicago Tribune* reporters Jeff Lyon and Peter Gorner tell it in *Altered Fates,* their authoritative book about the ex vivo story, "To this day, Anderson never talks about [the case] without thanking Dusty Miller, without whose packaging cells the experiment would never have taken place." But despite Culver's and Miller's close collaboration in the treatment, and Miller's seminal work to make it a reality, Anderson and his NIH colleagues Michael Blaese and Steven Rosenberg excluded Miller and Culver when they applied for a broad patent on the gene therapy technique.

For his part, Culver was stunned to notice the patent application one day in 1990 in a pile of papers Anderson had asked him to take to Blaese. As Culver told Lyon and Gorner, he had never seen a patent application, so he decided to flip through it. As he describes it, "I wound up in shock." The names on the patent applications were those of Anderson, Blaese, and Rosenberg. As Culver complains, incredulous, "They did this even though they had attached all my data on the back."

Miller, who now heads a research laboratory at the Fred Hutchinson Cancer Center in Seattle, explains that he and Culver pursued the matter legally because they had contributed the key technology involved. But despite the assessment of teams of attorneys at two separate law firms that Miller and

Culver should be listed as coinventors on the ex vivo patent, the matter stalled in arbitration at NIH. Miller says that the only remaining choice for him and Culver would be to sue NIH, but for the moment both are reluctant to take such a step. "My research is still government-funded," Miller says. "I don't think it would help a grantee to be suing NIH."

Today Miller and Anderson coexist within the esoteric, fast-growing scientific world of gene therapy research, and both sit on the editorial board of the new field's leading journal. But relations are strained. Anderson says it was lawyers who cut his colleagues out of the ex vivo patent, but Miller, who vows he will never collaborate with Anderson again, remains bitter that he was not named on the patent.

Miller's story is more than a simple case of sour grapes. It represents a bona fide difference of philosophy. "The important point here is that this is based on the work of many, many people — to sell that away into private hands — it's just not right," he says. "I think it is really a lot like the government selling the right to mine our national park land to private companies." Should Miller in fact win rights to the ex vivo patent, he says that, through the Fred Hutchinson Cancer Center, he would insist upon a nonexclusive licensing arrangement, allowing all researchers to have access to the technology. "The rights to this broad and powerful technology should be made available to everyone," Miller says, a view that is part of a time-honored tradition in medicine, from the development of anesthesia to Jonas Salk's polio vaccine.

But Miller's approach differs dramatically from that taken by the principals in charge of the patent on ex vivo human gene therapy. After a twisted path — and one that has been extremely lucrative to several private parties — ownership rights to all ex vivo human gene therapy are now controlled by a huge Swiss drug company. Miller minces no words in his assessment: "I think the public's been ripped off here," he says.

Ever since Congress opened the door in 1980 for research institutions to administer the patents they receive on publicly funded research conducted within their walls, the phenomenon of publicly funded experiments leading to private proprietary rights has become commonplace. The goal of the noteworthy legislation, the Bayh-Dole Act (passed in 1980), was to spur so-called technology transfer, the process of bringing the fruits of basic research to the marketplace. While the legislation's intent was laudable, the outcome has often been problematic, as shown by the case of the ex vivo patent.

During the time when DeSilva was receiving gene therapy, Anderson also headed the advisory board of Genetic Therapy, Incorporated (GTI), a small biomedical firm in Gaithersburg, Maryland. At Anderson's behest, GTI had entered into a cooperative research agreement with NIH to help fund his research and, eventually, the human gene therapy trial. When it came time to assign licensing rights to the broad patent the team had received, GTI was lavishly rewarded for its support, winning an exclusive license to the patent at no charge. As a result, anyone hoping to commercialize an ex vivo gene therapy in the United States had to negotiate a sublicense from the company.

Needless to say, James Barrett, president and chief executive of GTI, was pleased when he got the news. Barrett said he believed that the patent would be especially valuable because it would firm up the company's legal rights to gene therapies under development for patients with breast cancer, AIDS, and Gaucher's disease. And, Barrett hoped, it would "foster collaborations between his company and other gene therapy investigators and larger drug makers."

The interest of "larger drug makers" in GTI's patent portfolio materialized more quickly and dramatically than even Barrett could have hoped. Just three months after the gene therapy patent was issued in 1995, the giant Swiss pharmaceutical firm Sandoz bought GTI (which had no products or profits to speak of) for $295 million. In a scant four years, GTI, a firm that had started with $2.5 million in venture capital, had become a key asset in a multibillion-dollar conglomerate's strategy to lock up ownership rights to the emerging field of human gene therapy.

Meanwhile Ciba-Geigy, a rival Swiss conglomerate, was pursuing a similar strategy. It had just purchased a large stake in Chiron Corporation of Emeryville, California, which had its own strong patent portfolio in technology related to human gene therapy. Both Sandoz and Ciba-Geigy were speculating on a grand scale that this technology would soon spawn a lucrative new health care market.

How big was the potential market? About four thousand diseases are caused by inborn damage to one gene. The most common is cystic fibrosis. Cancer, AIDS, and many other illnesses are believed to result from malfunctions in one or more genes. As of 1998, thousands of patients are participating in gene therapy trials, and the majority of the experiments involve ex vivo techniques. Although no gene therapy product is being sold on the market today, the first are expected by the year 2000, including at least ten cancer gene therapies. By the turn of the century, industry watchers estimate the gene therapy market will top $2 billion. By the year 2010, the Federal Trade Commission estimates that the market for all gene therapy products will reach upward of $45 billion.

Announced in early 1996, the $63 billion merger of Sandoz and Ciba-Geigy created a Swiss-based multinational company on a truly staggering scale. The new firm, called Novartis, now stands as the world's number-one agrochemical corporation, second largest seed firm, third largest pharmaceutical firm, and fourth largest veterinary medicine company.

Many people are dismayed at the prospect of this corporate monolith. Says Miller, "If you have a gene therapy you want to bring to market, you have to go and negotiate with Novartis. They can say, 'Yeah, we're interested,' or they can say whatever they want. They are in a very strong position. Here's a group that claims patent rights to all of ex vivo gene therapy and they're blocking other people from advancing the field." Many observers worry about the power the exclusive license gives the company. As the University of Pittsburgh's Joseph Glorioso explains, in cases like Novartis's control over the ex vivo patent, "the company will be able to pick and choose winners in a very young field, and therefore to shape its development."

The prospect of the Novartis merger also drew the attention of the U.S. Federal Trade Commission (FTC). To do business in the United States, the huge new firm had to file papers that established that it didn't violate the nation's antitrust laws. The difficult question the FTC faced was this: what should the government do about an enormous foreign-based corporation designed in significant part to secure a private monopoly on the development of a promising and potentially lifesaving category of medical treatments that had been largely developed in the United States? Central to the FTC's concern was that Novartis, through its ownership stakes in GTI and Chiron, would be one of "only a tiny handful of entities capable of commercially developing a broad range of gene therapy products."

"Given the combination of the Ciba and Sandoz patent portfolios," the FTC ruled, and the twelve-year entry period required to bring gene therapy products to market, "it is extremely unlikely that any other firm would be able to enter the market to replace competition lost through the merger." While other firms might be technically capable of competing in the research and development of such products, the FTC said, they lacked the intellectual property rights for commercialization that this merger would put exclusively in the hands of Novartis.

Despite such antitrust concerns, the Federal Trade Commission, after some debate, allowed the deal to go forward, with a number of caveats. Chief among them was a mandate that in the field of human gene therapy, Novartis would have to license its patents, even to competitors. The FTC ruled that "in the broad gene therapy market," Novartis would be required "to grant licenses at low royalties to the 'Anderson ex vivo patent' . . . to any entity that requests such a license." Without such an arrangement, the FTC contended, an unduly strong patent portfolio in human gene therapy "would be solely under the merged firm's control."

Just as Congress had acted to forestall the worst excesses of medical procedure patents in 1996, the Federal Trade Commission had undertaken the most specific and direct exception it could to avoid the worst excesses of the Novartis gene therapy monopoly. To many critics, the limited regulatory action did not go nearly far enough. For instance, concerning the ruling that licenses must be made available at "reasonable rates," a number of critics asked, What is reasonable? Novartis might demand a payment of $10 million up front and a percentage of sales after that, a fee that might be reasonable for a very big firm, but would not be for a small biomedical start-up company.

As Dusty Miller explains, small companies have to consider the possibility of patent battles especially carefully. It is dangerous for the company and for investors if a firm doesn't have clear legal rights to the products and processes. And they might decide they are simply not going to take this risk. "For a company like Novartis, though, it is a different story," Miller says. "Fighting against Novartis is like going up against a Goliath. They can spend $100 million without even really thinking about it. And they can happily defer and delay until smaller players are forced to give up."

For the moment, it is unclear what Novartis will charge. Because of outstanding competing claims, the U.S. Patent Office has recently placed the Anderson ex vivo patent in a bureaucratic probationary limbo called "interference." Although it took six years to award the patent in the first place, the Patent Office is reviewing anew competing claims made at the time of filing that might affect the breadth of the ex vivo patent. Like so many similar tangled tales of disputed terrain on the intellectual property frontier, the case threatens to drag on for years, siphoning off funds and hampering the field's development. The effects of such unremitting legal wrangling can be felt not just in boardrooms but in the laboratories themselves.

When Business Rules

Even French Anderson, a staunch advocate of the commercialization of human gene therapy, has publicly wrung his hands over privatization in the biomedical field. In *Human Gene Therapy,* a journal he edits, Anderson complains that companies are forgetting about shared basic research goals in their rush to bring products to market. He writes, "I emphatically argue that they [private funders] must begin to support academic investigators to carry out gene therapy clinical protocols that have a strong basis in science but not in marketability."

Meanwhile, Steven Rosenberg, Anderson's colleague and fellow titleholder of the ex vivo patent, has expressed reservations of a different kind about changes in the biomedical field. Writing in the *New England Journal of Medicine,* Rosenberg, now chief surgeon and leading investigator at the National Cancer Institute, decries mounting secrecy in medical research: "In an era of decreased federal funding, interest by biomedical companies can be useful." Too often, however, the companies demand secrecy or confidentiality agreements, "because control of information from the results provides a competitive edge." For Rosenberg, the price of such secrecy is too high. Communication among researchers suffers, he warns, when "the rules of business precede the rules of science"; colleagues become unwilling to share their data.

Rosenberg supplies a handful of vignettes to underscore his concern. When cancer specialists met at NIH not long ago to discuss experimental human cancer vaccines, he recounts, an organizer stipulated that all new information presented at the meeting be kept confidential. In another instance, a scientist called him to say that he had "hot" data related to Rosenberg's research in cancer immunotherapy. When the two met to discuss a possible collaboration, however, the scientist began by asking Rosenberg to promise to keep the data secret. As Rosenberg recounts, "I refused and explained that I would not withhold from colleagues information that could be useful in their research. He would not agree. He left and I have not heard from him since."

Rosenberg has urged his colleagues to refuse to keep information confidential and refuse to sign any agreements that limit the open transfer of information. He says that more open discussion of the problem is needed in universities, hospitals, and professional meetings. And he offers a cautionary historical tale, likening the current situation to an infamous one in the seventeenth century, when the Chamberlen brothers, renowned doctors in Germany, developed the obstetrical forceps, a technology of great value in difficult childbirths. Despite the device's lifesaving capabilities, the Chamberlen family kept it secret for three generations, growing rich and famous for their obstetric work while women and children throughout Europe died in childbirth. Four generations after the invention of the forceps, the Chamberlen family finally sold the secret and it was made public. As Rosenberg charges, "Few would argue that such behavior was moral."

Today a related problem is occurring as private companies routinely fail to disclose, even in published papers, the composition of the biological substances they use called vectors and reagents. As Miller notes, GTI, now a division of Novartis, is notorious for this practice, routinely "explaining" the outcomes of its experiment as a result of the company's "proprietary" vectors. "An academic paper should say exactly how you've done the experiments. Other scientists need to know this information to make the paper useful," he says. "But companies don't want to let people know about gene therapy vectors — if other researchers ask, the company will not tell them." To make matters considerably worse, private biomedical firms routinely work with vectors and reagents given to them by academic researchers, then slightly alter the signatures by changing or adding a short nucleotide — essentially stamping them differently — so they can stake a proprietary claim on them. In the case of GTI, Miller says, the public was cheated when the company was sold for almost $300 million. "But if they continue over the next few years to stake proprietary claims to gene therapy vectors currently available in the public domain, the public will pay far more dearly."

The dearth of vocal advocates of the public's stake in both clinical health care and medical research helps foster a land-grab spirit on the intellectual property frontier. Unfortunately, the outcome of this approach does more than drain public coffers; ultimately it has an adverse effect on the health care the world's people will receive. In modern replays of the Chamberlen family's greed, private proprietary rights are frequently pitted squarely against public health needs.

In just one of numerous ongoing cases, for instance, many cancer patients are now being denied a promising treatment approved by the Food and Drug Administration (FDA) because Baxter International, a large pharmaceutical and health care firm, claims to own a broad license on the technology related to a particular antibody that can be used in bone marrow transplants for patients with breast cancer and lymphoma, among other diseases. Even though Baxter does not have a comparable FDA-approved product on the market yet, the company has legally blocked a small competitor — CellPro, in Bothell, Washington — from marketing its treatment while Baxter tries to bring its version through the lengthy FDA development and approval process.

The legal case drew national attention when CellPro developed a therapy to treat its fifty-year-old CEO, Rick Murdock, who was diagnosed in 1996 with a deadly type of lymphoma. But a fight over CellPro's right to the technology had already been brewing for two years. In 1995, CellPro won a favorable jury verdict in a patent case, but the judge threw out the verdict and Baxter won an injunction preventing CellPro from offering its lifesaving treatment to dying patients.

Andrew Yeager, director of bone marrow transplant programs at Emory University, where physicians have been using the CellPro treatment with some success as a last-ditch effort to save the lives of children suffering from acute leukemia, lamented to the *Seattle Times,* "It's unfortunate that these sorts of things in corporate America can threaten therapeutic clinical trials and potentially life-saving therapies."

More than three dozen members of Congress, the American Cancer Society, several patient advocacy groups, former U.S. senator Birch Bayh, Jr., and a former White House counsel, Lloyd Cutler, made an appeal on CellPro's behalf to Department of Health and Human Services chief Donna Shalala. They asked her to exercise the government's right to intervene in patent cases that stem from publicly funded research in "extraordinary cases" where the dispute threatens public health. "CellPro's request is simple. While the court case is allowed to run its course, an FDA-approved product must remain on the market, available to any and all cancer patients who need it," Bayh and Cutler wrote Shalala.

The government denied the request. NIH Director Harold Varmus, who made the ruling, said he was convinced the courts would ensure that no one was denied treatment. Varmus was undoubtedly swayed by the many influential voices arrayed against government intervention, such as Stanford University's President Gerhard Casper. Casper wrote to Varmus that intervention would set a precedent that "would pose a grave threat to university-industry partnerships" and even "put into jeopardy the kind of investments needed today to take medical discoveries through the lengthy processes necessary to bring them to the public."

Amid such powerful private interests and legalistic posturing, it is worth casting back a generation to see how differently these issues can be handled. In 1954, when Jonas Salk developed a polio vaccine, his funder, the March of Dimes, prohibited patenting or receipt of royalties on the results of its research projects. The notion of Salk individually owning rights to the discovery never came up. Nor did the idea of limiting licensing of the technology in an effort to personally control the direction of future research in the area. When Edward R. Murrow, the renowned television commentator of the day, asked, "Who will control the new pharmaceutical?" Salk replied that the discovery belonged to the public. "There is no patent," he said. "Could you patent the sun?"

Clearly, Salk's view is currently out of favor.

A New Test for the System

Shortly after President Clinton signed the medical procedure patent provision into law, another prominent case attracted widespread attention and made the front page of the *New York Times*. The case involves a common prenatal test.

Each year nearly every pregnant woman in the United States is given a blood test to screen for genetic defects in the unborn fetus. By measuring the concentration of several substances in the pregnant woman's blood, the so-called multiple-marker blood screen can warn a woman that her baby may have a genetic defect such as Down syndrome, which causes retardation. One of the substances measured in this way is human chorionic gonadotropin, or HCG, a hormone that women produce in the days after conception.

Since the mid-1960s, medical researchers have actively studied HCG and its role in building up the placenta. In 1989 a researcher named Mark Bogart was awarded a patent based on an observation he had made about HCG: that elevated levels of the hormone can signal the presence of Down syndrome in a fetus.

Bogart, whose work was done in 1986 at the University of San Diego, didn't create a new device to obtain his patent. He merely observed a connection between the level of HCG and the likelihood of Down syndrome — and recognized the potential use of this correlation in a diagnostic test. Nor did his observation by itself result in the multiple-marker blood screen, since his was only one of three separate observations that make up today's most commonly administered test. Nonetheless, combined with measurements of other factors in the blood, Bogart's research did open the door for the development of an inexpensive diagnostic test that alerts doctors to potential problems that might warrant more accurate and invasive tests.

Bogart received U.S. Patent No. 4,874,693, affording him monopoly protection over a "method for assessing placental dysfunction." Now he is turning his patent into dollars. Bogart claims the patent entitles him to a royalty of $3 to $9 every time a lab administers the multiple-marker test. He has made good on his threat to sue labs, doctors' offices, and health maintenance organizations that refuse to pay.

Many are paying. According to Andrew Dhuey, Bogart's lawyer, laboratories owned by SmithKline Beecham are now paying Bogart royalties in excess of $1 million per year. For example, the Arizona Institute for Genetics and Fetal Medicine agreed to Bogart's royalty demand covering all future screening tests, as well as paying $90,000 in royalties for tests conducted over the past six years.

Bogart's intellectual and financial claims have inspired outrage in some quarters of the medical community. As Arnold Relman, former editor of the *New England Journal of Medicine*, commented, "To claim private ownership rights over natural phenomena, the nature of disease, or human biology is a restriction of intellectual freedom that will stifle medical research. It is grotesque."

Critics like Relman stress the fact that Bogart did not invent anything in the traditional sense. Rather, he was granted a patent as a result of his noticing a natural bodily function that has probably always been present. As Robert Merges, a law professor and patent expert at the University of California at Berkeley, notes, the Patent Office has always denied patents on laws of nature. The Bogart patent, Merges says, "does seem a little like patenting Newton's law of gravity."

As Merges explains, it is so expensive to fight an infringement suit that "these patents have a tendency to terrorize" those accused of infringement. Consequently, some of the nation's most venerable hospitals, medical firms, and testing laboratories have followed their lawyers' advice and reluctantly paid Bogart's royalty.

Kaiser Permanente in California, the country's largest nonprofit hospital chain and health maintenance organization, is an exception. It has not agreed to pay and has challenged Bogart in court. Mitchell Sugarman, Kaiser's director of technology assessment, says the HMO has contested the claim because of what it implies for its patients and the broader medical community. Kaiser expects to spend well over $1 million to litigate this case, more than it would have paid in royalties. But Sugarman contends there is a "moral argument" to be won.

Joining Kaiser is a consortium of medical professional groups, including the American College of Medical Genetics, the American Medical Association, and the American College of Obstetrics and Gynecology. The group has offered Kaiser financial aid and free expert assistance. The organizations formed the consortium because in their view the Bogart case represents "a dangerous attack on the availability of an important diagnostic test for pregnant women and on public health policy in this country," says Michael Watson, the consortium's leader and vice president of the American College of Medical Genetics.

Nevertheless, because of Bogart's willingness to sue those he claims are infringing his intellectual property rights, Laurence Demers, president of the American Association of Clinical Chemistry, says many hospitals and labs are considering halting their use of the prenatal test. "Most people are really outraged by this whole thing," Demers says, "because they feel like it is extortion."

That feeling is all the more understandable, given the letter Bogart's lawyer sent to labs around the country that offer the screening test. The letter says the labs are faced with a "pivotal decision": they can either fight Bogart's patent infringement claim, an option Dhuey brands as "doomed to failure," or they can "pay that same money for a patent license which protects your lab while others fight that uphill battle for you." Should a lab challenge the patent in any way, Dhuey warns, "Biomedical Patent Management Corporation will sue your lab to recover for the last six years of your lab's patent infringement and seek an injunction against your continuing to use the patented procedure."

If Bogart gets the $3 to $9 royalty every time a lab administers a triple-marker test, as he is demanding, he could earn as much as $100 million in royalties from hospitals, laboratories, and medical research institutions over the patent's life. Experts in prenatal testing say the added costs could use up or exceed insurance reimbursement levels. But Bogart counters, "Why should they be allowed to make large profits, but those of us who provided the test not be allowed to make any?"

As *Bogart v. Kaiser Permanente* heads to court in California, it raises serious questions about whether anyone should be allowed to patent a function of the human body, or a medical method or procedure. As in most patent infringement cases, however, the court will not tackle those broad questions. Instead, the case will likely turn on narrow legal questions. Much of Kaiser's case, for example, will center on the argument that the currently used multiple-marker prenatal test is only distantly related to the original research Bogart patented.

The provision Congress enacted in 1996 prevents holders of certain types of patents on medical procedures from collecting royalties from health care providers. But the provision applies to new patents and therefore has no effect on Bogart's claim. Even if the law did apply, legal experts are divided on whether the Bogart patent — or others like it — would fall within the act's precise language. Greg Ganske, the Iowa Republican largely responsible for the bill's passage, said that patents like Bogart's may signal a need to strengthen the law, in particular to cover previously issued patents. "Maybe with examples like this we will need to go back at this again, " he said.

As Mitchell Sugarman at Kaiser notes, while Bogart may well deserve some credit for discovering — or at least sharpening — our realization of the correlation between HCG and the probability of Down syndrome, "his financially motivated claim does nothing to benefit medical science."

For his part, Jack Singer says he was "deeply troubled" when he heard of the Bogart case. "Patients, the medical profession, and society all benefit from the long-standing culture of free exchange of medical knowledge," he says. Only patent owners and their lawyers benefit from carving up medical knowledge into privately held parcels. Singer's personal experience with a patent-infringement case has forced him to think deeply about the issue. "The foundation of good medical care is interdependent." From the "discovery, evolution, and early evaluation of technological advances to the rapid dissemination of improved techniques," Singer says, "a culture of free and open exchange improves the quality of patient care and greatly benefits the health and welfare of our society."

Unfortunately, however, as numerous legal battles in biomedicine attest, in a knowledge-based economy this free exchange cannot be taken for granted. If it is not actively championed, the time-honored ethic of sharing medical knowledge could become as rare as a doctor's house call.

Perhaps Mark Evans, professor of obstetrics and gynecology at Hutzel Hospital in Detroit, expressed it best when he condemned Bogart's patent claim on

ABC News. "If this patent is enforced, it will have serious consequences to the health care of women in this country. At the very least, it will significantly raise health care costs. And women who otherwise would have been able to identify pregnancies that had serious problems might not be able to identify them."

"I believe in capitalism and rewarding discoveries," Evans proclaimed, "but there has to be a point at which social responsibility takes over greed."

5
Softwars

NOVEMBER 1993, LAS VEGAS, NEVADA. Technophiles flock to Comdex, the computer industry's biggest trade show, for the thrill of the new. In a field renowned for a breakneck pace of technological innovation, the enormous, bustling event has showcased a cavalcade of cutting-edge products from laptop computers to virtual-reality software. Vendors here do everything they can to ensure that their newly minted high-tech products delight, dazzle, and even shock. In 1993, though, one of the biggest shocks of the event came from neither hardware nor software but from one small company's announcement at a carefully staged press conference.

Stanley Frank, then president of a California-based firm called Compton's New Media, used the Comdex show to focus industry attention on the broad intellectual property rights his company had recently secured. Compton's had already made a name for itself by publishing one of the earliest and most successful CD-ROMs — *Compton's Interactive Encyclopedia* — a so-called multimedia reference disk that combined sound, graphics, and text. People in the industry had closely tracked the growing popularity of the CD-ROM encyclopedia. What they didn't know, however, is that in 1989, when Compton's first launched the product, the company had also filed for a patent covering all multimedia software. And just months before the Comdex show, the U.S. Patent Office had granted the company the sweeping ownership rights it had sought.

To Stanley Frank, Comdex offered the perfect venue at which to gloat about the monopoly the government had handed his company. "We invented multimedia," he told a packed crowd at the Comdex press conference. His company's patent substantiated the claim.

In this case the sanctioned title of "inventor" had powerful implications indeed. Compton's patent did not cover specific programming language. Rather, in some forty-one separate claims, it gave the company exclusive ownership rights over any multimedia database software that allowed users to search simultaneously for text, graphics, and sounds. According to the patent's broad language, Compton's was the exclusive owner of the "process and concept" of so-called retrieval technology in multimedia databases.

Frank couched his initial announcement of the patent in magnanimous terms. "We simply want the public to recognize Compton's New Media as the pioneer in this industry," he said, adding that the company hoped to "promote a standard that can be used by every developer." But as the press conference wore on, it quickly became clear that recognition was only part of what Frank sought. As the sole owner of a patented technology, the company could legally exact royalties from anyone else who wanted to use it. As Frank put it, Compton's was determined to be "compensated for the investments we have made to make multimedia a reality for developers and end users." In the terms the company laid out, all other multimedia CD-ROM manufacturers would have to either pay between 1 and 3 percent of their revenues to Compton's or negotiate a joint venture with the company if they wanted to sell multimedia products over the course of the seventeen-year lifetime of the patent.

While there was no overt mention of it at Comdex, Compton's message also carried an implicit threat. Any multimedia publisher that didn't comply with the company's terms could face a costly patent infringement lawsuit. And if this unlucky firm lost the case, it would have to pay even more dearly: patent infringers can be subject to treble damages — a punitive payment three times greater than any economic harm Compton's might allege in the case.

To competitors of Compton's New Media, the prospect of a broad patent in their promising new field posed a serious and inescapable problem. The search tools Compton's now claimed to own were a basic feature of virtually every multimedia product. With scores of multimedia CD-ROM titles already on the market — many of them on display at Comdex — the patent not surprisingly fostered a good deal of outrage. The larger implications were particularly troubling. If other companies secured similarly broad patents, the proliferating royalty demands would seriously erode the potential for profits in the multimedia field.

Philip Dodds, executive director of the Interactive Multimedia Association, a trade group representing 260 companies from various industries, called

Compton's announcement "one of the most serious developments" the industry had ever faced. Dodds complained that the patent was too broadly defined and covered a search process already widely seen as nonproprietary. Rob Lippincott, president of another trade group, the Multimedia Industry Association, voiced his concerns more bluntly. He called Compton's patent "a 41-count snow job."

At Compton's press conference, Norm Bastin, the company's executive vice president and general manager, contended that the company had earned exclusive ownership rights because "this sort of search system was unique" at the time it filed for the patent. The company's patent lawyers had evidently convinced the U.S. government of Bastin's contention. But now Compton's faced a tougher audience. Assembled experts asserted that the techniques the patent described were widely used before the 1989 filing date. Some noted, for instance, that the technique for indexing and searching multimedia databases had been explored originally at the Xerox Corporation's Palo Alto Research Center — Xerox PARC — almost two decades before Compton's made its claims.

But whether or not Compton's originated the concept, most competitors objected to the breadth of the claim. "Patenting multimedia is like patenting the English language," said Robert Carberry, president of Fireworks Partners, a New York–based IBM affiliate, when he heard the news. As one former Microsoft executive put it, the absurdity of one company owning exclusive rights to something as broad as multimedia technology made most competitors want to jump on Compton's "like a herd of elephants on a mambo snake."

But the ridicule and anger did not alter the fact of the patent's existence nor the fact that it represented a powerful weapon, valid until proved otherwise. And attempting to overturn a patent through the legal system can be an extremely costly and unpredictable venture. The gravity of the situation was quickly apparent to players in the multimedia industry. Joe Clark, CEO of Videodiscovery, a Seattle-based company that publishes science-oriented multimedia titles, captured the prevailing sentiment when he likened the news at Comdex to a major earthquake. To competitors like him, Clark said, Compton's announcement registered "7.3 on the Richter scale."

Clark was particularly sensitive to the potential damage of a broad patent like Compton's because his company had already gone to great expense to fight off two other patent claims. The new threat seemed too much to bear. "My big worry is that there's no end in sight," Clark said. "We don't know what to expect tomorrow, next week, or next month."

Eventually, as the aftershocks of Compton's earthquake subsided, industry representatives began to shift their attention away from the firm at the epicenter, directing their ire instead at the U.S. Patent Office. How could the examiners have overlooked the fact that the concept of multimedia had been in circulation long before Compton's claimed to have invented it? The Compton's announcement focused the attention of the entire software field on the proliferation of software patents that threatened to wreak havoc on the industry.

To most programmers, Compton's patent represented much more than a mistake by an inexperienced patent examiner, and it concerned more than an assessment of what the Patent Office terms "prior art." Even if Compton's could unequivocally establish that it was the first to develop a multimedia database, the patent's breadth opened the door for a crippling tangle of similar claims, with royalty demands that could choke off the field's prized pace of innovation. At the heart of the Compton's debacle was a multibillion-dollar question: What exclusive ownership claims, if any, were viable over concepts that the entire software field needed to use to develop new products? Could the industry, or the patent system, ever hope to reliably distinguish between specific innovations that might merit individual recompense and broad chokeholds over shared concepts?

Silicon Valley's Lament

The Compton's saga posed the classic and increasingly familiar dilemma of broad conceptual ownership claims threatening to distort and stifle development in high-tech fields. It also presented precisely the kind of nightmare that some in the software industry had predicted for years. The fact is, software has proven exceptionally troublesome to the U.S. legal system since it first appeared in the 1960s.

On the one hand, software code can alter a computer's functioning so significantly that it can be seen as creating an entirely new machine. And, following this line of reasoning, the invention of new machines in the United States has always been allowed the protection of patents. On the other hand, though, software code itself is not a machine. It is made up of strings of instructions, something like the recipes in a cookbook. Historically, the U.S.

legal system has tended to treat instruction sets — even useful and lucrative ones — as forms of expression protected solely by copyright law.

The distinction is far from trivial. The different kinds of ownership systems imply widely different rights. Most notably copyright allows practitioners more latitude. A composer's symphony may be protected by copyright but the constituent parts — the musical notes, chords, and conventions of time and meter — cannot normally be privately owned. Under the copyright regime these constituent parts are held by all of us in common, reused from one piece of music to the next.

Despite its utilitarian nature, software shares many of the features of a piece of music or a written work. It is made up of a progression of subroutines that are roughly equivalent to the specific steps or directions outlined in a musical score or a recipe. And, as programmers are keenly aware, these subroutines are widely repeated from one program to another. Programmers need to have unfettered access to the language of software in order to write new programs.

Recognizing the need to maintain widespread access to these shared features, the Patent Office simply refused to grant patents on software programs in the early days of the computer age. The Supreme Court ruled in 1972 that software fell into the category of "mental processes" and held that its logical steps had to be preserved in the public domain "as they are basic tools of scientific and technological work." Copyright therefore remained the only legal method available to programmers seeking to protect their developments, and the industry thrived and blossomed with that protection.

But in a global economy increasingly fixated on the value of knowledge assets, some companies sought tighter private control over their particular advances in software design. A key shift occurred in a 1981 Supreme Court case involving a company whose application for a patent on a system for manufacturing rubber included a software program to control the temperature throughout the process. In *Diamond v. Diehr,* the Court ruled by a narrow margin that the inclusion of a software program in a patent application should not automatically disqualify it from consideration by the Patent Office.

Diamond v. Diehr pushed the door to software patents slightly ajar. But the Patent Office's liberal interpretation of the ruling soon flung it wide open. Before long the agency was awash in a flood of applications for patents on software programs. By the early 1990s software patents were one of the fastest-growing sectors of the U.S. patent system, being issued at nearly three times the rate of other kinds of patents.

Many in the software field began to complain that the agency's patent examiners were ill prepared to rule on the profusion of patents in this fast-paced field. They charged that the Patent Office's decisions often seemed capricious and that many of the patents it issued seemed unreasonably broad. Patents were beginning to divide the emerging software field into arbitrary monopolies that squelched newcomers and new developments alike.

Mounting tensions came to a head in the aftermath of the multimedia patent issued to Compton's New Media. With the Compton's patent and a growing number of high-profile intellectual property lawsuits in the software field, many contended that urgent action was required to address the situation. Bruce Lehman, commissioner of the Patent and Trademark Office, stepped into the breach early in 1994. He took the highly unusual step of second-guessing his own examiners, personally calling for the agency to reexamine Compton's patent application. In addition, he scheduled an unprecedented hearing in Silicon Valley to hear the industry's views on what to do about proliferating software patents to defuse the sense of imminent crisis.

By the standards of Washington, D.C., the hearing was anything but typical. First of all, the packed hall in the San Jose Convention Center, normally home to annual meetings of chiropractors and Rotarians, lacked the staid authority of Capitol Hill. More notable, though, was the distinctive character of the crowd. The youthful software programmers and entrepreneurial executives tended to mistrust the federal government. In their views and demeanor, they would prove to be far different from the lawyers and lobbyists Lehman and his fellow Patent Office officials normally dealt with.

Opening the meeting, Lehman laid out his agenda. The best kind of patent system, he said to the hushed crowd, is clear, understood by everyone, and requires little litigation. "Our concern is that we're not quite seeing that kind of patent system, particularly in the software-related inventions area." Over the next two days, Lehman and his team would be pummeled with criticism and complaints. Only the patent lawyers who testified were uniformly sanguine about the direction of development in the field; the overwhelming majority of programmers and executives had little good to say about the way the rules of intellectual property were affecting their industry.

Douglas Brotz, principal scientist at Adobe Systems, was one of the first to testify. Adobe, based in Mountain View, California, was well respected in the field for its widely used PostScript programming language and desktop publishing tools. Brotz got right to the point. "The emergence of patents on software has hurt Adobe and the industry," he contended, relaying a tale of woe that would be repeated many times before the two-day hearing was over. "Resources that could have been used to further innovation have been diverted to the patent problem," he complained. "Engineers and scientists such as myself, who could have been creating new software, instead are focusing on analyzing patents, applying for patents, and preparing defenses. Revenues are being sunk into legal costs instead of into research and development."

As Brotz recounted, Adobe had already been sued for patent infringement by a competitor in a lengthy and costly lawsuit. Although Adobe had won the initial case and the appeal, it had cost the company more than $4.5 million in legal fees and expenses. Employees had spent thousands of work hours on the case, with the chairman of the board devoting a month just to appear at the trial. "I myself have spent over 3,500 hours of my time — that's equivalent to almost two years of working time," Brotz said.

Consequently, Brotz called for the Patent Office to return to its policy of refusing to patent software. "I take this position as the creator of software and as the beneficiary of the rewards that innovative software can bring in the marketplace," he said. A world without software patents hadn't stopped his company from creating new programs, nor had it deterred the venture capitalists who helped Adobe with early investment. Perhaps software patents might be justified if they brought some other kind of benefit to the field, Brotz said, "but I see none." He concluded that conferring monopoly positions "will promote stagnation rather than increased innovation. When companies turn from competing by offering the best products to earning money by the threat of patent litigation, we will see our best hope for job creation in this country disappear."

Jim Warren, a director of the California-based computer firm Autodesk, had a similar message for Lehman and the panelists from the Patent Office, but he was more direct in his exasperation and anger. Warren's credentials and wealth of experience gave his testimony added authority: he was the founding president of the Microcomputer Industry Trade Association, the founder of the field's first subscription newspaper, *InfoWorld*, and had helped start a number of successful software firms. He also held graduate degrees in medical information science, mathematics, and statistics, as well as computer engineering. Warren did not mince words. "There is absolutely no evidence whatsoever — not a single iota — that software patents have promoted or will promote progress," he said.

Facing the panel directly and referring only occasionally to the prepared remarks before him, Warren testified that of the thousands of programmers he had known over the past quarter century, not a single one ever said they developed a program because they wanted a monopoly on it. But in the current climate, Warren said, Autodesk, like most other firms, felt forced to try to secure patents on its software techniques to defend itself against others who might try to monopolize them. All the effort to secure patents, he said, represented "an infuriating waste of our technical talent and financial resources made necessary only by the lawyers' invention of software patents."

According to Warren's testimony, Autodesk had faced no fewer than seventeen patent infringement claims over the past several years. The company had spent well over a million dollars defending itself. And, Warren lamented, "millions more are certain to pour down the bottomless patent pit." Warren said his company was fortunate: it had the financial and technical resources to rebuff baseless ownership claims made against it. But he was not happy about having to devote such significant resources to the cause. "Your office has issued at least sixteen patents that we have successfully rebutted," Warren charged, the frustration evident in his voice. "We never paid a penny in these attempted extortions that your office assisted. But they have caused an enormous waste of resources that could better be invested in useful innovation."

Warren's complaints were also voiced by larger firms. Jerry Baker, senior vice president of Oracle Corporation, recommended that patent protection be eliminated for computer software. On behalf of Oracle, a fast-growing company boasting $1.5 billion in revenues and employing more than 11,000 people worldwide, Baker asserted that "software patents are failing to achieve the Constitutional mandate of promoting innovation and indeed are having a chilling effect on innovative activity in our industry."

Baker said the rapid pace of development in the software field put the patent system's seventeen-year monopoly "completely out of context with industry reality." More importantly, though, he stressed the difference between software and other types of inventions. "Software seldom includes substantial leaps in technology," he said, "but rather consists of adept combinations of several ideas." As Baker put it, "Whether a software program is a good one does not generally depend as much on the newness of each specific technique, but instead depends on how well these are incorporated into the unique combination of known algorithms and methods. Patents simply should not protect such a technology."

Like Warren, Baker said that his company felt coerced into participating in a patent system it believed to be fundamentally flawed. "Our engineers and patent counsel have advised me that it may be virtually impossible to develop a complicated software product today without infringing numerous broad existing patents." As a result, "Oracle has selectively been applying for patents which will present the best opportunities for cross-licensing between Oracle and other companies who may allege patent infringement."

Nearly all of the programmers who testified over the course of the hearing voiced similar complaints. Each seemed to have a favorite analogy for what was happening. One likened the situation to individual schools suing one another, claiming exclusive ownership of techniques like long division. Another compared it to carpenters having to pay a royalty every time they picked up a tool.

Tim Boyle, head of a consortium of software companies that included Compton's New Media, had his own analogy. Undaunted by the prospect of alienating one of his prominent constituent firms, Boyle urged the Patent Office officials to allow fundamental concepts like "multimedia" to remain in the public domain. As he quipped, "How would theater have developed if the concept of 'plot' were owned by someone? William Shakespeare never could have afforded a license."

Late in the proceedings, someone challenged Lehman with the fact that the only people testifying in favor of the patent system were lawyers. The point was not lost on Lehman. "There is no question that the lawyers seem to be very much in favor of patent protection," he reflected. When a programmer griped that the panel was made up exclusively of lawyers, Lehman deflected the evident hostility in the audience with humor. "Sorry, we run the world," he retorted. "Julius Caesar was a lawyer, you know. The pharaoh was a lawyer. You can't get away from that."

The clash of cultures represented at the Silicon Valley hearings was perhaps most clearly highlighted in the testimony of Richard Stallman. Stallman, a respected programmer and recipient of a MacArthur Foundation "genius" award, was well known in the field as a champion of shareware — software programs disseminated freely by their creators, who receive voluntary nominal payments from users. And he had come prepared with a few crowd-pleasing missives.

"The Supreme Court has ruled that no one can patent an algorithm or other law of nature," Stallman said, "but skilled patent lawyers have been tricking the Patent Office into regularly doing precisely this in the software field." Stallman testified that a colleague of his, curious to test the limits of the system, had applied for and won a patent on Kirchoff's Law — an 1845 scientific theory holding that the electric current flowing into a junction equals the current flowing out. Stallman said his colleague sought the patent not to reap any financial benefit but to confirm his suspicion of serious deficiencies in the system. "If the Patent Office couldn't understand electricity after a century," Stallman ventured amid murmurs of approval from the audience, "how can we expect it to understand software in another decade or two?"

Like many of the other speakers, Stallman emphasized the difference between software and other types of patented inventions. "In some fields, like pharmaceuticals, one patent goes with one product," Stallman said. "Software is the extreme opposite. A typical patent covers many dissimilar programs, and even an innovative program is likely to infringe many patents. That's because a substantial program must combine a large number of different techniques and implement many features."

To illustrate his point, Stallman produced a voluminous, unwieldy printout of a computer program he had written with several colleagues, explaining that the program is in use on more than a million computers, including those of the U.S. Air Force and major companies like Intel and Motorola. "Just a few lines of code can be enough to infringe a patent, and this compiler has ten thousand pages," Stallman said, gesturing to the document. "How many patents does it infringe? I don't know. Nobody does. Perhaps you can read the code and tell me?" he challenged Lehman. His guerrilla theater made his point effectively, drawing hoots of laughter from the crowd.

"An invalid patent is a dangerous weapon," Stallman explained forcefully. "Defending a patent suit typically costs a million dollars and the outcome depends mostly on legal technicalities." But he also underscored the fact that the problem went beyond the question of technical validity. Suppose the Patent Office stopped making mistakes and issued no more invalid patents, he said. "Suppose that it is the year 2010 and you're a software developer. You want to write a program combining 200 patentable techniques. Suppose 15 of them are new; you might patent those. Suppose 120 of them were known before 1990; those would not be patented any longer. That leaves 65 techniques probably patented by others, more than enough to make the project infeasible. This is the gridlock we are headed for.

"A decade ago, the field of software functioned without patents, and it produced innovations such as Windows, virtual reality, spreadsheets, and networks. And because of the absence of patents, programmers could develop software using these innovations," Stallman said. "We did not ask for the change that was imposed on us. There is no doubt that software patents tie us in knots. If there's no clear and vital public need to tie us up in bureaucracy, untie us and let us get back to work."

As the hearing wore on, it appeared that Silicon Valley's lament was making an impression on Lehman. When twenty-nine-year-old Ted Lemon, a software engineer at Network Computer Devices, reviewed a series of broad, harassing patent infringement claims leveled against his company, the tale seemed to particularly catch Lehman's attention.

"The essence of the problem, then," Lehman summarized, is "that there was a patent issued that didn't meet the test of patentability."

"Right," said Lemon.

"And now, in effect it's being used to extort money out of people, and they just buy into the extortion scheme and then they pay up rather than solve it.

"You know," Lehman reflected, "it reminds me a little bit of the old thrillers that you used to see on television when I was a kid about the Mafia holding up the candy store, and people would let that happen, you know, getting protection money out of them." He paused for an awkward moment. "Maybe that makes me the vigorous prosecutor; maybe that's my role to do that," he mused in an odd moment of self-reflection that the audience greeted with a sustained round of applause.

Signs of Trouble

Within a year of the Silicon Valley hearing, the U.S. Patent Office reexamined Compton's patent and rejected every one of its forty-one claims, citing "new evidence" of prior art that had come to light. To the relief of many, the

Compton's debacle had been successfully defused. Despite Lehman's intervention, however, the overarching question about the viability of software patents has grown ever more confounding and ominous since Compton's patent first surfaced in 1993.

After the Silicon Valley hearing, the Patent Office issued new internal guidelines for patenting software, reiterating its intention to avoid patenting most mathematical algorithms. In addition, the agency modified the rules to allow third parties to participate in the process of reexamining questionable software patents. These changes were warmly welcomed by the industry. But neither altered the fact that, legally speaking, the prevalence of patented software subroutines and techniques made every software manufacturer in the industry vulnerable to attack.

Early in the 1990s the intellectual property problems in the software field led Mitch Kapor, formerly of Lotus Corporation and now a principal of a cyberspace think tank, the Electronic Frontier Foundation, to predict an impending meltdown in the industry due to proliferating lawsuits. Referring to the industrial disaster that struck the Union Carbide plant in Bhopal, India, in 1984, he warned of a "Bhopal of software patents." Kapor's meltdown has yet to occur, but signs of trouble abound.

Take, for example, the case of Vern Blanchard. As president of a small, San Diego–based company called American Multi-Systems, Blanchard wrote a software program in 1993 for use in professional bingo halls and was starting to market his system, along with custom-built tables and personal computers, when one of his competitors filed suit for patent infringement. In Blanchard's case it didn't matter that he finally triumphed in the courts. His competitor won an initial court injunction that prevented American Multi-Systems from bringing its product to market and scared off prospective clients. The tactic saddled Blanchard's company with legal fees that put it out of business and more than $100,000 in debt. The case is but one illustration of how vulnerable small companies can be to aggressive patent litigation.

Blanchard's tale underscores one dramatic kind of hazard, but many less catastrophic warning signs have surfaced over the years. As early as 1990, for instance, the threat of a patent infringement suit forced the makers of the XyWrite word processing program, XyQuest, to send out a revised version of their software to those who had purchased it. XyQuest had independently developed the automatic correction and abbreviation features included in their XyWrite III Plus program. But company executives decided they couldn't risk the cost of a lengthy court case, even though they believed they had proof of prior art that ultimately could have invalidated the patent. As a result, XyQuest simply removed the features and offered its confused and unhappy customers a "downgrade": a version that actually took away features that users were already enjoying. The case is a small but telling example of the needless, unproductive downside of software patents.

There is also growing evidence of heightened secrecy and mistrust in the software development field. Conference organizers, for example, report that it is difficult to sign up speakers on technical matters. As Russell Brand, the organizer of one well-known state-of-the-art technical conference, explained several years ago, "More than half of the speakers that I approached said they couldn't speak this year because of patent-related restrictions placed upon them by their company's corporate counsel." As a result, "It's going to be another two or three years to find out what they are doing, and so everyone working in that same field isn't going to be able to build on that research as quickly." Brand says he never had such problems when the conference series started in the mid-1980s.

These kinds of problems are magnified as the software field continues to gain prominence within the U.S. economy. A 1997 study ranked the $100 billion software industry as the nation's fastest-growing sector and the third-largest manufacturing industry after automobiles and electronics. An unprecedented 11,500 software patents were issued in 1997, and the number has been rising by roughly 30 percent every year for the last decade. At this rate, according to one estimate, there will be some 80,000 software patents in force by the year 2000.

From this byzantine maze of intellectual property claims, two distinctly problematic types of patents emerge: those monopolizing the algorithms present in many different software packages and those claiming exclusive rights to overly broad concepts.

Computer programmers particularly disdain the annoyance of patents on the small chunks of computer code called algorithms, the discrete mathematical strings of rules that allow a computer to perform such basic tasks as moving a cursor around the screen or alphabetically sorting a list — both of which have been patented. Technically, algorithms are not patentable, but patent lawyers continue to cleverly disguise them as computer-implemented processes.

Should a programmer wish to develop a program that includes language to generate footnotes, he or she would risk violating a patent, no. 4,648,067. The same would be true for techniques to compare documents (patent no. 4,807,182), to allow multiple users to access a database at the same time (patent no. 5,642,503), or to determine the amount of time a slide or transparency is viewed in a presentation (patent no. 5,642,430). In 1997, Xerox even won a patent for a method of keyword searching in a document that has arguably been in place for nearly three decades. These kinds of patents on specific subroutines are causing even greater headaches as companies stake claims over the computer processes that control how information moves across the Internet and how it is selected and organized by users.

The prevalence of broad conceptual claims is just as problematic. Some cover business practices that would never be patentable if they were not carried out on a computer. Firms have won patents on methods for processing data in a banking terminal, on methods for processing inventories and filling out order forms, for evaluating risk in the insurance industry, even for exchanging money from one currency to another. In one of the earliest examples, Merrill Lynch patented a so-called case management account system in 1982. The software procedure does nothing more, essentially, than move investment funds among different types of accounts. Acknowledging that the system would not be patentable if executed with pencil and paper, the U.S. courts nevertheless upheld the patent because the tasks were accomplished on a computer.

The Crystal City Test

Greg Aharonian, the publisher of *Patnews,* an Internet patent news service, is an expert on software development who investigates the validity of software patents on behalf of firms involved in lawsuits. He claims that 50 to 70 percent of the software patents issued can't pass what he calls "the Crystal City test." That is to say, if U.S. patent examiners ventured beyond the walls of their headquarters in Crystal City, Virginia, to query average software programmers, they would find that the techniques they are patenting are already widely known and used. Assuming that the current rate of software patenting continues, Aharonian predicts that by the end of the decade some 40,000 questionable patents will plague the industry.

Consumers are likely to feel the impact of this proliferation of ownership claims in three distinct ways. First, we will probably pay higher prices for software as companies are required to pay other firms for licenses when they create new products. In addition, because the system favors larger firms with extensive patent portfolios, we will be less likely to find products from smaller, fledgling companies. Perhaps most important, though, the dubious nature of many of these claims needlessly threatens to undermine the vibrant pace of innovation in this field for all players.

Why do so many questionable patents continue to slip through the system? Some answers can be found in ten bland office buildings in the sterile, concrete landscape of Crystal City, across the Potomac River from the nation's capital. A visit to the U.S. Patent and Trademark headquarters here finds an agency that, despite its age and pedigree, must surely rank as one of the oddest and least-known branches of the federal government. In these buildings some 1,800 well-paid federal examiners (aided by thousands of support staff) process and pass judgment on roughly 175,000 patent applications annually.

The first thing to impress a visitor is the scale of the enterprise. The patent library holds some 23 million documents pertaining to the nearly 6 million patents issued to date. Leafing through these stacks, the extent to which patents pervade our lives becomes clear. No mundane gadget has been forgotten. Before reaching the breakfast table, almost all Americans make use of household items covered by scores of U.S. patents — from toothbrushes and shampoos to showerheads, not to mention alarm clocks, light fixtures, door hinges, and sneakers.

The team responsible for issuing patents on "computer software-related inventions" is called Group 2300, a section of examiners that exudes a feeling of fast-paced change. Having grown tremendously over the past five years, Group 2300 is now one of the Patent Office's largest departments, employing some 200 examiners. The group's document handling room, where software patent applications are processed, looks like a good-sized post office. Alan MacDonald, a supervisor and senior examiner in Group 2300, leads a brisk tour of the area. As he explains, the department used to be divided by subject matter into two separate units. Now it has fifteen. Some 15,000 patents are currently pending here. And the cases are often notoriously complex. MacDonald points out shelves upon shelves of pending applications, many in folders that are more than a foot thick.

By the standards of the largely traditional and antiquated Patent Office,

Group 2300 is high-tech. To do its job, the group has access to a new computerized database, with side-by-side screens that allow examiners to retrieve both the graphics and text of any previous patent granted. Nonetheless, the job of granting patents in the software area is plagued by the field's tremendous fluidity. Software routines are developed, built upon, reused, and slightly altered in many different kinds of programs. Employees shift from one firm to another, bringing with them certain styles and ways of writing code.

To make matters considerably more difficult, software developers leave little or no paper trail. Group 2300 might as well be charged with trying to uncover who first spoke a word or coined a phrase. In software development, unlike other fields, there are few peer-reviewed journals to help the Patent Office determine its important criterion: who invented something first. The problem was prophetically anticipated in the report of a 1966 presidential commission, which noted, "The patent office now cannot examine applications for [software] programs because of a lack of a classification technique and the requisite search files. Even if these were available, reliable searches would not be feasible or economic because of the tremendous volume of prior art being generated. Without this search, the patenting of programs would be tantamount to mere registration and the presumption of validity would be all but nonexistent."

As MacDonald laments, "In this field especially, there is no magic book to help you track key advances." The patent system's vaunted technology includes no books at all, and texts are actually a place where widely accepted basic algorithms are set forth. In an effort to compensate for this shortcoming, Group 2300 recently added a powerful computer system that can sift through stacks of CD-ROMs for written records of prior art that might pertain to a software patent application.

A number of bureaucratic realities compound the agency's difficulties. For instance, MacDonald, whose specialty is artificial intelligence and speech signal processing, is an extreme rarity for the length of his tenure in the group. Since he was hired, eleven years — and more than a million patents — ago, Group 2300 has continued to expand steadily, in spite of federal cutbacks. But the group still faces a dire shortage of experienced examiners, especially in light of the Patent Office's requirement that an examiner must have worked in the agency for seven years to have full "signatory authority" to actually grant a patent. The bottleneck is worsened by the addition of new units (each requiring at least one supervisor) and the drain of some staff leaving to work in the private sector. Group 2300 has also seen some of its most experienced supervisors promoted to the Patent Office's Board of Appeals to deal with the multitude of software lawsuits. Given this climate, it is not hard to understand how the agency can issue patents that are widely ridiculed in the industry.

Charles Van Horn, deputy commissioner of the U.S. Patent Office, insists that such cases are the exception, not the rule. All the Patent Office needs to combat its "tremendous problem retaining examiners," especially in fields like software and biotechnology, is better funding. Once the examiners get proficient enough, he explains, they tend to be snatched up at much higher salaries by firms desperate for their expertise. With more experienced examiners, Van Horn says, the Patent Office might be better able to avoid issuing overly broad patents.

Van Horn's recommendations, while no doubt constructive, do not address the system's fundamental problems. One feature that is often overlooked, for instance, is that the patent process is an iterative one. Overall the Patent Office ultimately grants patents to some 60 percent of all applicants, but only 10 percent of the applications are approved on their initial presentation. As Alan MacDonald explains, the system of granting a patent is one of give and take between agency and applicant. From the perspective of the shrewd applicant, therefore, a patent accepted on the first presentation is likely to mean that the inventor didn't claim enough — much like the seller of a home who immediately is offered the asking price. Much of the patent attorney's job comes in trying to claim as much as possible — to construe an applicant's patent very broadly in order to win the greatest protection.

After all, the attorneys argue, broad and vague patents have been around since the first known patent on a technological innovation, granted to Filippo Brunelleschi, the architect of Florence's remarkable cathedral, in 1421. Brunelleschi claimed that he had invented a new means of conveying goods up the Arno River, but he vowed to remain intentionally vague on details unless the state promised to keep others from copying his design. At the time, the city-states of Italy were locked in an internecine struggle for wealth and power, and Florence agreed to Brunelleschi's conditions, affording him the right to exclude all new means of transport on the Arno for three years. If the system has been viable in the past, today's analysts argue, it can continue to be in the future.

In a similar vein, some insiders emphasize the cyclical nature of patent litigation and its dependence on the maturity of a given industry. Because patents are granted based on precedent, or prior art, it stands to reason that new fields offer particularly big potential rewards. For example, patent expert Robert Merges views the current problems in high-tech patents as a relatively common phenomenon in the infancy of industries, akin to growing pains, which will subside as the field matures. "A biotechnology start-up firm," Merges says, "is nothing more than its technology. A lot of money is at stake in its patents as companies vie for position." Ultimately, though, he contends, the rival claims settle out and firms go back to business as usual. For instance, in the automotive industry, General Motors might win an extremely broad patent on catalytic converters, but Chrysler probably won't be too worried about it. As he puts it, "They probably have a broad patent on transmission systems already." According to this view, the present rise in software patent disputes can be expected to subside with time.

Unfortunately, though, these views underestimate how different the software industry is from previous manufacturing sectors. Because software is intangible and even ephemeral, these firms will never be "anything more than their technology," to borrow Merges's term. And they are all drawing from the same pool, creating software out of the same evolving programming languages. Examples of this problem abound. At Group 2300, for instance, MacDonald notes a recent case in which seven separate companies in the artificial intelligence field applied for almost identical patents. All of these applicants had invented essentially the same thing in the same six-month period. "Cases like that may cause headaches," MacDonald says, "but they also keep my job exciting."

What he sees as excitement, though, can also be read as a profound obstacle. Writ large, it is clearly evident in one of the biggest ongoing fights in the software industry today: the gigantic battle between Microsoft and Sun Microsystems over the now widely accepted programming language called Java.

While the battle over Java is not simply a fight over software patents, it illustrates several important problems with the current system. Sun Microsystems developed Java and made the programming language available to anyone who wished to use it. Large firms who wanted to include Java in their products were required to license it, but Sun took a magnanimous approach to the licensing process, anxious to find wide acceptance for the new language. Designed to run on any machine (to "write once, run anywhere," as Sun puts it),

Java allows programs to reside anywhere on the World Wide Web. Many companies are now developing Java applications; according to John Heilemann, reporting in the *New Yorker,* IBM has more people working with Java than Sun does — some 2,400 employees in 1998.

In March 1996, Microsoft signed a deal with Sun to license Java for use as a feature on some versions of its Windows operating system. But the union was ill fated. In October 1997, Sun sued Microsoft for trademark infringement, false advertising, breach of contract, and unfair competition. Microsoft countersued, also alleging breach of contract, among other charges. At the root of the case is Sun's contention that Microsoft is trying to undermine Java's "write once, run anywhere" nonexclusive quality by developing applications in the language that work only in conjunction with Windows.

Scott McNealy, CEO of Sun Microsystems, explained his company's position in testimony before the Senate Judiciary Committee in March 1998: Microsoft's "exclusionary business practices," he charged, give it an undue measure of control over "the written and spoken language of the digital age." He says he won't tolerate Microsoft's appropriation of Java in a proprietary manner. As McNealy told the Senate committee, the only thing he would rather own than Microsoft's Windows operating software would be exclusive rights to the English language. Then, as McNealy put it, "I could charge you two hundred and forty-nine dollars for the right to speak it, and I could charge you an upgrade fee when I added new letters."

McNealy highlighted a profound problem. Languages evolve incrementally, but they must be widely and freely accessible to be useful. The current environment of increasingly expansive private ownership in the software field undermines this cornerstone of shared access. And the problems will grow in coming years unless the system is corrected.

What is to come?

Kapor's dire scenario, in which software companies self-destruct in a tangle of litigation, still lurks as an ever-present danger. Like many programmers, Stallman says that he lives in fear of "a horrible disaster happening any day" — an intellectual property lawsuit against him that he will be powerless to defend, or one that tears the entire field asunder.

Currently, though, an eerie standoff has developed. With major software firms building sizable arsenals of intellectual property as a deterrent, needless barriers to innovation are rising, as are fear and confusion on the part of even the larger players. The situation is especially poignant because so few of those involved respect the validity of the legal protections most companies have garnered.

The current situation tends to favor larger companies. As Adobe's Douglas Brotz explains, "The expensive patent process protects large, methodical corporations that can afford to apply for scores of patents much more than it protects the poorly capitalized lone inventor, and when that inventor tries to protect his invention he may well find that those large corporations can ruin his own business with their large software patent portfolios."

Take IBM, for example, the industry leader in software patents, with a new one nearly every day. The company's position is further strengthened by the leverage its thousands of patents provide. In 1997, IBM used its patent portfolio to secure at least fifty-two cross-licensing agreements with other firms. While almost all observers agree that cross-licensing is better for the industry than widespread litigation, it favors those firms who can bring larger and more powerful patent portfolios to the negotiating table.

In addition, the growth of cross-licensing does little to alter the lingering threat that some companies will exploit their intellectual holdings as a competitive tool. The fact is, in what *Fortune* magazine dubbed several years ago the "sue-the-bastards approach," a growing number of firms are intentionally using patents to squeeze competitors.

The kind of problem exemplified by the Compton's case has not gone away. To the contrary, it has become institutionalized. For a sense of the kind of regular eruptions we can expect in the future, consider the shock wave a virtually unknown company called E-data sent through the industry in the spring of 1996.

Three Men and a Patent

E-data made its public debut in the spring of 1996. The tiny, three-person firm sent a letter to 75,000 separate companies warning that if these firms were conducting business over the Internet, they were likely infringing E-data's broad patent on Internet commerce.

E-data outlined its stance in a series of advertisements that supplemented the company's mail campaign. The message was simple. The ads, featuring a picture of a carrot and a stick, read, "Your choice." The "carrot" E-data offered companies was amnesty: companies that signed up immediately for a licensing arrangement — involving a relatively modest annual fee of $5,000 to $50,000, depending on the company's revenues — would be exempted from royalties on past infringement of E-data's alleged intellectual property. The "stick," of course, was the threat of a patent infringement lawsuit.

How did E-data ever obtain a patent on something as broad as financial transactions on the Internet? The story begins in 1985, when computer programmer Charles Freeny won a patent by outlining a system in which products are purchased on line and delivered electronically. It was a simple but prescient idea. Anticipating a world of networked computers, Freeny sought to patent on-line transactions such as downloading music or magazine articles on demand. With little prior art to be found, the U.S. patent examiners granted Freeny Patent No. 4,528,643, entitled a "System for Reproducing Information in Material Objects at a Point of Sale Location."

Despite Freeny's farsightedness, though, he could not have envisioned the patent's universal applicability to the mushrooming on-line commerce of the World Wide Web. In 1989, unsure what his patent might ever be worth, Freeny sold it to an entrepreneur for $200,000. (Less than ten years later, E-data publicly estimated that it expected to collect licensing fees on some $20 billion of digital transactions by the year 2000.)

Freeny's patent resurfaced with a vengeance in 1994 when it was bought by an enterprising businessman named Arnold Freilich. Seeing the patent's lucrative potential, Freilich moved fast. With two partners, he built the corporate shell for E-data on the back of a tiny company called Interactive Gift Express, Inc., which, something like an FTD florist, distributed gift packages of stuffed animals, bathroom soaps, and specialty foods.

Establishing himself as president and CEO of E-data, Freilich hired patent lawyer David Fink, whom he fondly describes as "the pit bull of patent infringement." Early on, Freilich and his partners at E-data were lampooned in the trade press as "three men and a patent," but the joking quickly dissipated as a number of financial analysts recognized that E-data's strategy, if its patent held up in court, might well succeed in exacting royalties from every company that wanted to buy and sell in cyberspace.

In response to E-data's "carrot and stick" campaign, a number of firms, including IBM, Adobe, Intermind Corporation, and Kidsoft, Inc., agreed to license the patent rather than risk fighting it. In royalty negotiations with IBM in 1996, one of the less discreet negotiators on the IBM team speculated publicly that E-data's patent, if valid, could be worth billions. In a heady three weeks that year, E-data's stock price soared from $1.63 to $11 per share.

Meanwhile, though, the vast majority of the 75,000 firms E-data had contacted by mail failed to sign up for licenses, opting instead to see whether E-data's patent would hold up in court. As a result, in the summer of 1996, E-data began to make good on its threat to sue infringers. So far the company has filed lawsuits against forty-one companies, including the Internet provider CompuServe, the financial firm Dun and Bradstreet, various software makers, including Broderbund and Intuit, and publishers McGraw-Hill and Ziff-Davis. These firms have been forced to sink millions of dollars into fighting the case in court.

"Part of our marketing strategy was to sue everybody and get noticed," Freilich said. "Well, we went ahead and sued, and everyone now knows that we're very serious about defending our claims."

"We don't want to affect anyone's ability to do business," he says, adding that he is only trying to provide his company's shareholders with "a fair return on their investment." As he puts it, "I hate being called a leech, but such is life."

Needless to say, few of the 75,000 companies targeted by E-data have much nice to say about the company. Its tactics have been derided as "patently offensive," a "nuisance," and even an "abomination." It doesn't help matters that the firm exists merely to exploit a patent.

Nonetheless, compared to the stir caused just a few years earlier by Compton's multimedia patent, most players seem surprisingly resigned to a sanctioned system in which companies extort money from each other by claiming to own absurdly broad and seemingly obvious conceptions. As Stewart Baker, a Washington, D.C., attorney who's fighting E-data's patent on behalf of several businesses, says, "We have clients who have said that at my level of business right now it is cheaper just to pay the license than to even ask my lawyers to examine what the defenses might be."

Much of the cynicism comes from the myopic limitations of the court system in dealing with the underlying problem presented by patents like E-data's. The legal battle against its claims, for example, hinges on an interpretation of Freeny's initial conception. The lawyers for CompuServe and many other defendants seek to declaw the claim by arguing that the "point of sale" described in the broad language of Freeny's patent actually envisioned a kind of on-line cash register for dispensing information at retail outlets. Whether this line of argument will win in court remains to be seen. The patent language is so vague that even if the court rules for a narrowed interpretation, the hair-splitting, semantic legal battle will fall far short of addressing the problem in future cases.

Regardless of Freeny's precise vision, if he even was the first person to ever think of on-line commerce (a highly dubious proposition by the early 1980s), the broader question is whether anyone should be able to own exclusive rights to such a sweeping concept — precisely the same question that caused the uproar in 1994 over Compton's multimedia patent. The question will undoubtedly cause numerous battles in the years to come, yet it is one that is outside the jurisdiction of the courts. It must be dealt with on a broader level of policy or legislation.

As Stallman puts it, for instance, "Two kinds of patents hurt programmers: valid patents and invalid patents. Even a valid patent can and will obstruct software development."

The E-data case is currently still pending. Recent developments in the trial indicate that the judge might be receptive to the defendants' arguments for a narrow interpretation of the Freeny patent, but the outcome remains unclear. A definitive verdict, including the appeals process, is likely to be several years and many millions of dollars away.

Close watchers of the software field warn that we will see many more such cases emerge from the formidable backlog of pending software patent claims. A youthful software firm called Open Market, for instance, announced in 1998 that it has been awarded several patents covering significant aspects of on-line commerce. In particular, the company was awarded U.S. Patent No. 5,715,314 for the concept of electronic shopping carts, a popular feature currently found at many Web sites to allow on-line customers an easy way to accumulate items while they browse the site. In addition, its U.S. Patent No. 5,724,424 covers secure, real-time payment using credit and debit cards over the Internet. It is one of the broadest Internet payment patents yet granted, with a filing date of December 1993.

According to patent watcher Greg Aharonian, Open Market, like E-data before it, is likely to seek royalties from other companies involved in Internet commerce. And, Aharonian says, Open Market's patents may be even more difficult to contest than E-data's. "The company included a commendable number of prior art references in their patents," Aharonian explains. "There will be no easy knockout here to allow you to ignore letters from Open Market."

He notes, "I estimate that every month for the last three years at least one software patent has been issued that is comparable to the Freeny patent: potentially infringed by everyone on the Internet, with overly broad independent claims and little or no nonpatent prior art submitted. As people start making money on the Internet, you can fully expect these patents to start being asserted."

Certainly, the prospect of many more companies like E-data seeking to exact royalties from everyone in the industry is enough to give pause to almost any player in the software field. There is little clear justification for a state of affairs in which patent holders — especially those who have made no contribution to the field — can hold the entire industry hostage to dubious conceptual claims. It is simply another example of our inability so far to set limits on the scope of ownership claims in the emerging knowledge-based economy.

Because Aharonian's most lucrative contracts come as a consultant in software patent litigation, though, he sides with many attorneys in the field who welcome the prospect of proliferating lawsuits. "Let them all sue each other," Aharonian quips. "Software gridlock is Nirvana to a guy like me."

6
Soybean Dreams

DECEMBER 1990, CLAY COUNTY, IOWA. During the growing season, the Winterboer farm buzzes with life, an 800-acre patch in the northwest corner of the state, thickly carpeted with corn and soybeans. But one gray morning in winter, the season when Iowa's flat farmland looks its emptiest, the sheriff of Clay County drove out to Denny and Becky Winterboer's farm to deliver some bad news. As Becky remembers it, the sheriff, a part-time farmer himself, asked if he could bring the Winterboers his soybean crop to clean with their mechanized harvester, which strips away the pods and stems. But, he said, his visit was official: he had come to serve legal papers notifying the couple that they were being sued by Asgrow Seed Company, then a division of Upjohn.

"It just didn't seem possible," Becky recalls. "I was flabbergasted." As Denny puts it, "We had no idea how it could be; there had been no notification from the company, nothing about the matter was even in the wind." The settlement terms of the four-year legal debacle that followed prohibit the Winterboers from discussing the case in detail. But the legal arrangement can't shroud their residual bitterness. "We're pulling out of it now," Denny Winterboer says, "but if I can ever imagine what hell is like, I feel like I've been there, and it's not pretty." The Winterboers' crime, according to the Asgrow complaint the sheriff brought that day, was selling some of their soybean crop to their neighbors.

Since 1987 Denny and Becky Winterboer had engaged in the time-honored practice of so-called brown-bag sales. That is, in addition to growing soybeans to sell as food and livestock feed, the couple sold a percentage of their sizable crop to their neighbors for seed. Such sales are nothing new; Denny's family has farmed in Iowa for four generations, and neighboring farmers have sold each other seeds for far longer than that. And, whether the seeds are sold, traded, or just set aside, farmers have saved a portion of their crops for replanting since the dawn of agriculture. The practice lies at the heart of farming.

The problem in the Winterboers' case, however, was that their soybean plants grew from two varieties of seed purchased from Asgrow Seed. The company claimed proprietary rights over these seeds — *and future generations derived from them* — no matter who grew and harvested the crops.

The Winterboers made slightly more money selling their harvest for seed than for feed: by convention, soybeans sold for seed, especially the latest varieties, fetch a slightly higher price. Becky Winterboer describes their sideline as "farmers helping farmers." With some crops, like corn, varieties are specially bred for planting. These hybrid crops lose their growing power — what plant breeders call their hybrid vigor — in successive generations; if you plant corn kernels from a hybrid variety, you won't get a very good yield. But soybeans, a self-pollinating crop, are not hybrids. As a result, the soybeans the Winterboers sold their neighbors to sow as seeds were identical to the ones they had bought from Asgrow. From a practical standpoint, it made little difference to the Winterboers how their soybeans were used.

To the Asgrow Seed Company, however, the end use of the soybeans sold by the Winterboers was of crucial concern. Of course, the company wanted the Winterboers to grow — and to sell — soybeans from its seed. But by selling soybeans for their neighbors to plant, the company claimed, the Winterboers were stealing Asgrow's intellectual property: specifically registered seed varieties with the prosaic names A1937 and A2234. Not the least of Asgrow's concerns was that the Winterboers were selling their neighbors A1937 and A2234 seed for about half the price Asgrow charged.

Concerned about competition from the Winterboers, the multimillion-dollar seed division dispatched Robert Ness, a neighboring farmer and an acquaintance of Denny Winterboer's father, to call on the Winterboers and ask to buy some soybeans. Ness purchased twenty bags of each variety and promptly handed them over to Asgrow, where plant biologists verified that they were identical to A1937 and A2234 seeds.

The ensuing legal battle between Asgrow and the Winterboers raised such profoundly unresolved issues that the case eventually made its way to the U.S. Supreme Court. Legally, the Winterboers' case rested on the strength of a piece of the 1970 Plant Variety Protection Act known as the "farmer's exemption." Although the 1970 legislation strengthened the intellectual property rights of seed companies like Asgrow, Congress had explicitly stipulated that farmers could save proprietary seed to replant on their own land or sell to their neighbors, as long as both the buyer and seller were farmers whose primary farming occupation was the growing and selling of crops "for other than reproductive purposes."

The farmer's exemption reflected the age-old view that farmers, in the United States and around the world, play a role in developing new crop varieties and in maintaining genetic diversity in the varieties they grow. As long as seed companies could recoup their investments and thus would continue to develop productive and hardy new plant varieties, the federal government reasoned, it made sense to let farmers retain the right to preserve, trade in, and develop varieties of the crops they grew. Among the benefits of the scheme, a small but healthy trade in less popular seed varieties would help protect the nation's crops against widespread susceptibility to rusts, fungi, blights, and other diseases that occasionally devastate popular varieties.

Perhaps even more important, though, the farmer's exemption reinforced a long-held social convention. Historically, purveyors of agricultural goods like seeds or plants have not been able to control how growers used those goods. Someone can sell you an apple tree, for instance, but they have no claim to the apples you harvest in the future. Nor can they prevent you from planting seeds from the apples you harvest to grow more trees of your own.

In the 1980s, though, the application of genetic engineering techniques to plant breeding began to strain such long-held notions. Viewed for millennia as the means to a crop, seed is gradually coming to be seen as an embodiment of intellectual property, as a blueprint, in other words, that carries value of its own. Jack Kloppenburg, an agricultural sociologist at the University of Wisconsin at Madison, eloquently expresses the view that has come to predominate. As he puts it, "Now a seed is, in essence, a packet of genetic information, an envelope containing a DNA message." This information packet "becomes the nexus of control over the determination and shape of the entire crop production process." To be sure, the change has been in the making for several generations, but now the notion of the seed as embodied genetic information has become the explicit mission of companies like Asgrow. Today the firm even describes itself in its corporate literature as "a developer, producer and marketer of genetics." The fight between Asgrow and the Winterboers simply exposed the extent to which the expansion of intellectual property claims in agriculture now puts titleholders and growers at odds.

If the Winterboers' case spotlights the shrinking control growers have over the seeds they buy, an equally vexing question is where to draw the line on the type of ownership rights Asgrow asserted. The farming industry, which has traditionally placed a premium on a farmer's ability to utilize new techniques and equipment to increase crop yields, is now looking elsewhere for its newest gains, namely, to plant breeders' manipulations of the seeds themselves.

It is a familiar shift. In agriculture, as on many other high-tech fronts of the knowledge wars, a rise in the value of intellectual property assets presents a fierce challenge to the established rules. Farmers are still doing the farming. The issue is who will control the knowledge assets they plant in the ground.

Greatly complicating matters is the fact that plants grow. Gradually and naturally, as thousands of years of plant breeding have demonstrated, plants adapt to changing environments over successive generations. Even more clearly than with many kinds of industrial inventions, new seed varieties owe an immense debt to the stock from which they derive. Notwithstanding the power of the latest genetic engineering techniques to alter characteristics, no plant breeder, high-tech or low, starts from scratch in developing a new plant variety. The work, in a very real sense, is part of a global ecological, evolutionary web. In this sense, Garrison Wilkes, a biologist at the University of Massachusetts, has rightly noted, "The major food plants of the world are not owned by any one people but are quite literally a part of our human heritage from the past." The simple fact is that most companies that receive proprietary protection for new plant varieties today are, in their breeding programs, unabashedly appropriating genetic traits from the dwindling storehouse of shared public plant varieties.

As the Winterboers can attest, the private capture of formerly shared varieties is not just fodder for academics like Wilkes and Kloppenburg; it is a battle quietly raging on arable land around the world. Thus, for Asgrow, the Winterboer case is as much about protecting market share as it is about the principle of protecting intellectual property. And along these lines, it matters immensely that soybeans are big business.

Seeds of Change

Many Americans may think of soybeans as an exotic health food or as an ingredient vaguely associated with the soy sauce they pour on Chinese food.

The crop did originate in China some five thousand years ago, but the perception of its limited uses couldn't be more inaccurate. Soybeans are a major pillar of the world's agricultural trade, the United States' second largest crop after wheat, and the world's foremost source of protein and oil. According to the American Soybean Association, the United States grows slightly more than half of the world's $27 billion soybean crop, well outpacing the output of other large producers like Brazil, Argentina, and China.

Because a large portion of the soybean crop in this country is used for livestock and poultry feed, its value to the agricultural economy far exceeds even the $14.5 billion its harvest fetches in the United States: it is a key ingredient in a food chain that grosses some $60 to $70 billion annually. With such an enormous worldwide market, conglomerates like Du Pont, Monsanto, and American Cyanamid are keenly aware that relatively small improvements in crop yields can translate into huge profits. According to a U.S. government estimate, each 1 percent gain in productivity from germplasm introduced into an important crop such as soybeans can result in as much as a $1 billion economic boon. In essence, undergirding the Winterboers' legal battle is the deeper question of how to apportion that potential profit between growers and plant breeders.

The outcome of this battle has implications for all of us. When private firms gain exclusive control over the intellectual property needed to grow the food that we rely upon for survival, they will, in essence, own the future: dictating the direction and shape of agricultural development and, in a very real sense, determining the types of food that are available to us. In this scenario, we will pay doubly — in grocery store prices dictated by these seed monopolies and, perhaps more important, in narrowed choices about how to grow the food that sustains us all.

▪

Against the backdrop of this high-tech, multibillion-dollar farming enterprise, it is hard to imagine how differently seeds were viewed even a few generations ago. Back in the late 1920s, for instance, when Denny Winterboer's grandfather was farming in Iowa, there was no legal provision for owning a seed variety even if someone wanted to. In fact, the famous California plant breeder Luther Burbank, arguing for patent protection for new plant varieties in 1930, lamented before the U.S. House of Representatives that "a man can patent a mousetrap or copyright a nasty song, but if he gives to the world a new fruit that will add millions to the value of the earth's harvest, he will be fortunate if he is rewarded by so much as having his name associated with the result."

A generation before that, at the turn of the century, when Denny Winterboer's great-grandfather homesteaded on Iowa land, the still-youthful U.S. Department of Agriculture freely distributed more than 20 million packets of seeds to the nation's farmers. It was one of the department's central mandates, as envisioned by Congress in 1862 when it established the agency, to "propagate, and distribute among the people new and valuable seeds and plants." The success of farmers like Denny Winterboer's ancestors in selecting and breeding these crops quite literally built the agricultural base of the United States.

In a climate where seeds themselves were freely distributed, it was unimaginable that an incremental improvement in a crop's characteristics would be something valuable enough to go to court over. This was especially true for the lowly, undervalued soybean. In the early 1800s the hardy legumes were even used as ballast by clipper ships, whose crews tossed them overboard upon reaching the New World. North American farmers didn't grow soybeans at all until 1829, and the crop didn't really catch on until the early 1900s, when George Washington Carver touted their virtue as a valuable source of protein and oil.

It is worth noting that the soybean's place in U.S. agriculture was finally assured only through the remarkable efforts of an American plant breeder named William J. Morse. In 1929, Morse began a two-year voyage through China, gathering more than ten thousand soybean varieties to bring home. According to the American Soybean Association, the varieties Morse culled "laid the foundation for the rapid ascension of the U.S. as the world leader in soybean production." Viewed by today's standards, in other words, the modern-day soybean that sustains a multibillion-dollar harvest in this country — including the myriad privately held varieties like Asgrow's A1937 and A2234 — is largely derived from a brazen piracy of intellectual property on a vast scale.

For his part, Denny Winterboer warns that farmers will eventually rebel against companies' increasing "private control over the gene pool." But some, like Hope Shand, research director at the North Carolina–based Rural Advancement Fund International, point out that such rebellions are already sprouting in several parts of the world. In 1993, for instance, as the Winterboers' case reached the appeals stage, the GATT negotiations provoked some five hundred thousand farmers in India to demonstrate in protest of private ownership rights to seed varieties. In a related incident, an irate group of a thousand Indian farmers ransacked the Bangalore offices of the multinational Cargill Seed Corporation, burning the office's documents in a huge bonfire.

"Although we hear little about this controversy in the United States," Shand says, "the issue of control of seed varieties is extremely controversial in the developing world." She explains, "Indian farmers are angry because they don't want to pay royalties on seeds and other products that they believe were developed using their own genetic resources and knowledge." Testifying before Congress in 1994, Shand argued that government policymakers around the world "would do well to take a cue from the Indian farmers." Sooner or later, she warns, the public will "demand a broad societal review of intellectual property laws affecting biological products and processes."

Given her concern for the plight of farmers worldwide, it is easy to see why Shand would take a keen interest in the Winterboers' legal fight. When the case reached the Supreme Court, the Rural Advancement Fund International, with help from nine other public-interest groups, filed an amicus, or friend-of-the-court, brief on the Winterboers' behalf. As the brief explains, the case presents a challenge to farmers' rights to "save, use and sell seeds harvested from crops they have grown through dint of their toil, at their expense, and on their lands." Such rights, the brief argues, have "existed virtually unfettered since the dawn of civilization."

Presumably swayed by this perspective, as well as by the protections offered in the law's "farmer's exemption," the federal appeals court had ruled that the Winterboers were within their legal rights. Supreme Court Justice Stevens agreed, stating that the statute clearly indicated that "Congress wanted to allow any ordinary brown-bag sale from one farmer to another." After all, Justice Stevens wrote, such a straightforward reading of the provision was in keeping with the courts' time-honored resistance to restraints on private property. "The seed at issue is part of a crop planted and harvested by a farmer on his own property. Generally the owner of personal property — even a patented or copyrighted article — is free to dispose of that property as he sees fit."

But the Supreme Court majority saw the case differently. Ignoring the broader debate over the rights of farmers to their harvest, and overturning the judgment of the appeals court, the high court opted for a kind of semantic gymnastics, narrowly reinterpreting the wording of the farmer's exemption as it is spelled out in section 2543 of the Plant Variety Protection Act. Writing for the majority, Justice Antonin Scalia argued that the Winterboers' brown-bag sales had exceeded the intent of Congress when it established the farmer's exemption and that the couple had indeed illegally profited from the intellectual property of the Asgrow Seed Company.

After years of effort, expense, and turmoil as they shepherded their case through the U.S. court system, the Winterboers were, of course, deeply dismayed by the outcome. So were Shand and many others interested in retaining a vestige of the notion of a public domain in plant breeding. But the ultimate outcome of the case was overshadowed by a far more dramatic turn of events.

Shand's amicus brief had ended on an impassioned note, urging that "on a matter fraught with such important consequences for agricultural development and the relationship between farmers and suppliers, the seed industry should not be permitted to win in the courtroom what it has failed to secure through the legislative process." Ironically, the brief's resounding conclusion quickly became moot.

In August 1994, on the eve of oral arguments in the Winterboer case, Congress dealt what Shand calls "a crushing blow to farmers' rights," amending the Plant Variety Protection Act to remove the farmer's exemption altogether. The action was motivated in large part by a desire among U.S. legislators to strengthen international intellectual property provisions. But the combined impact of the court and congressional decisions was decisive: in the United States it is now expressly illegal for farmers to sell or save seeds from proprietary crop varieties without receiving permission from breeders and paying royalties.

Shand says she recalls just a few years ago hearing the chair of the House Agriculture Committee dismiss as a paranoid fantasy the idea that farmers would ever be prevented from saving their own seeds. In fact, few agricultural policymakers anticipated the pace of change in the field. "Over the past two decades, I have witnessed a remarkable expansion of power afforded to the people who breed plants and own plant varieties," Shand says. "Many of us have worried about these issues for some time, but I don't think any of us guessed we would see such sweeping changes happen so quickly." The underreported but dramatic legislative and legal changes that occurred over the course of the Winterboer case explicitly codified the expansion of power given to breeders of new plant varieties.

The Winterboer case marks a watershed, with the U.S. government now sanctioning a new form of sharecropping, as farmers essentially become renters of plant germplasm. "What we're seeing," Shand says, "is nothing short of a new kind of 'bioserfdom.' Only this time, instead of controlling the land, the new feudal lords — the large agrochemical firms — gain their power and wealth by owning the information contained within the new high-tech seed varieties."

Species Claims

By the mid-1990s a series of legal and legislative decisions had paved the way for plant breeders to own agricultural genetic information. But the monumental battles over ownership of this information have just barely begun. Already, Shand's "high-tech, feudal lords of agriculture" are making increasingly broad and audacious claims, including some over entire species of plants. And, with a flurry of mergers over the past decade, seed companies have consolidated to a mere handful of dominant firms. Already just ten to twenty multinational seed and plant biotechnology companies dominate the most profitable segments of the global seed trade. As these agrochemical firms clash, the intractable difficulties entailed in delineating one party's ownership claims from another's come to the fore, making Asgrow's claims in the Winterboer case seem downright paltry.

The problem is abundantly clear in the claims of a Wisconsin-based biotechnology firm called Agracetus. Using a novel method, Agracetus researchers were the first to successfully insert genes into the cells of soybean seeds so that the plants would have altered characteristics. Based on a single successful genetic manipulation, Agracetus won a 1994 European patent affording the company exclusive rights for the next two decades to *any and all* genetically altered soybeans, *created by theirs or any other method that might be developed.*

The Agracetus patent represents no small prize. By controlling all genetic engineering changes to an entire plant species, Agracetus won a powerful kind of dominion. Scores of laboratories around the world are currently working to genetically alter soybean varieties. And many leading thinkers in agriculture share the view of Agracetus's former vice president Russell Smestad that eventually "all crops will be transgenic," that is, their genetic makeup will be added to or otherwise modified to boost yields or offer such long-sought properties as resistance to pests, frost, even drought.

Geoffrey Hawtin, director-general of the International Plant Genetic Resources Institute, was one of many who were angered by news of the Agracetus patent. Hawtin strongly denounced it, fuming that "at a stroke of a pen, the research of countless farmers and scientists has potentially been negated in a single, legal act of economic hijack."

From the perspective of Agracetus, however, the ruling of the European Patent Office comes after many grueling years of research, allowing the company to begin securing a string of similar protections around the world for its research investment. Agracetus has already filed for nearly identical soybean patents in the United States, China, India, and Russia. If all goes according to plan, the company may even realize its goal of controlling exclusive worldwide rights to the next several generations of improved varieties of soybeans, corn, and cotton.

The case raises several hotly contested questions. Given that Agracetus was the first to successfully insert genes into soybeans, should the company be entitled to own the particular variety of soybean its researchers created? By pointing the way, should the firm be granted ownership rights for a limited period to the particular process it used to insert the genes? Or does finding one method of inserting genes into soybeans give Agracetus an exclusive claim to the notion of genetically manipulating this species in any way? These difficult questions remain unresolved but, astonishingly enough, the firm's European patent implies not some but *all* of these rights.

Disputes over issues like those raised in the Agracetus case are likely to continue around the world for many years to come. In this respect, the fight to control the knowledge of how to make a better soybean is instructive. Whether or not their claims hold up against legal objections from other parties, the story of how Agracetus came to own all transgenic soybeans provides a fascinating glimpse into the wild and wide-open frontier of intellectual property claims in the biological sciences. No part of the tale is straightforward, and no one's rights are clearly defined in the shifting nexus of international legal protections, fast-paced technological developments, and changing business alliances.

The story of Agracetus's soybean claims begins, quite literally, with a bang. Ning-Sun Yang, a researcher at Agracetus, distinctly remembers hearing it from a lab down the hall one morning in the spring of 1988. Dr. Yang's colleague Dennis McCabe, a brilliant and largely self-taught experimentalist, had gotten the team's "gene gun," as it has come to be known colloquially, to work. As McCabe shot genes into plant cells with the velocity of a bullet, the sound reverberated throughout the laboratory complex in southern Wisconsin. The present gene gun resembles a pistol, but the Agracetus prototype was more like a cannon, a refrigerator-sized apparatus that now resides in the Smithsonian Museum collection in Washington, D.C. Its initial explosion in McCabe's lab would hail a discovery with profound implications for both agriculture and ownership as the world heads into the twenty-first century.

Formed in 1981, Agracetus was a biotechnology start-up firm that found a niche specializing primarily in three key American crops: cotton, corn, and soybeans. Because of their biological makeup, all three of these crops were proving hard to manipulate with genetic engineering. A major thrust of Agracetus's research was to develop a system to insert genes into the cells of these crops. A plant breeder who could accomplish this could avoid the painstaking months or even years needed to alter the traits of the plant and its progeny through more traditional methods. Agracetus's solution was a gene gun that uses helium gas and an electrical charge to spray microscopic gold beads into the reproductive area of a plant cell. The beads are coated with genes — fragments of DNA that govern particular characteristics in the developing plant, from hardiness and disease resistance to size and drought tolerance.

Leaving the gun at supersonic speed, the tiny gold pellets blast through the cell's outer membranes by brute force but, remarkably, seem not to damage the cell or the delicate structures inside. As Yang explains, gold is the most inert and nontoxic heavy metal known. And the gold beads are infinitesimal. If the cell were the size of an average office, Yang says, the DNA-coated gold bullets would be the size of a tennis ball. When the genetic material lodges in the cell, it introduces into the organism — a soybean seed, for instance — new characteristics.

As the researchers toiled to get their new gadget to work, Agracetus coworkers dubbed Ning-Sun Yang, Dennis McCabe, and Paul Christou "the three musketeers." Using their varied backgrounds in microbiology, plant genetics, and engineering, the three worked in earnest on the gene-gun problem for several years in the mid-1980s. Yang still remembers fondly the celebrations over the gene gun's success. Many, such as his close friend and colleague Dick

Burges, a biologist at the University of Wisconsin, told him that with the gene-gun technique Agracetus "was sitting on a gold mine."

As the first team to successfully blast genes into soybean cells, the three musketeers of Agracetus had good cause to celebrate. In assessing the strength of the company's patent claim, though, it is important to note that although they contributed to the refinement of a working gene gun, they did not invent the concept. Like virtually all technological development, their work built on a robust foundation of previous scientific invention. As Yang freely admits, the group hatched the idea for a gene gun after reading a pathbreaking article by a team at Cornell University. There a group of scientists led by John Sanford and Edward Wolf had built a prototype of what they called a "particle gun" and had written several articles describing the results and outlining their novel theory of so-called particle bombardment to introduce genes into plant or animal cells.

In Yang's history of events, his group simply "outcompeted" the Cornell researchers. But the reality is not nearly so simple. Agracetus made some important contributions toward the achievement of a working gene gun: its researchers switched the metal used for the microscopic beads to gold from the tungsten used by the Cornell group; they refined the unit's helium-gas propulsion system. But the fact is, the Cornell group's technology received a seminal patent on "particle bombardment technology" in 1984, four years before Agracetus won U.S. Patent No. 5,120,657 for its version of the gene gun. Agracetus *was* the world's first company to successfully transfer foreign genes into soybean cells that could be successfully regenerated into whole plants. But the breakthrough built upon the contributions of many other researchers — a fact that is highly relevant, given the company's sweeping ownership claims. The three musketeers at Agracetus jumped excitedly on the landmark work of a university research team that, in turn, derived a large measure of its knowledge base from nearly two decades of public funding of recombinant DNA research in the United States and abroad.

Given the litigious nature of this emerging field, the two groups' competing claims on the actual gene-gun devices might well have been resolved in court, but the companies opted instead to settle their differences in the relative privacy of the corporate boardroom. In 1989 Du Pont bought an exclusive license to use Cornell's patented particle gun in exchange for $2.8 million in royalties and research support. In so doing, Du Pont leased to others the only commercially available model of a gene gun. Agracetus, on the other hand, opted not to license its gene gun to any other companies or researchers, offering instead to have Agracetus researchers perform custom transformations of plant varieties for high-paying customers. In May 1992, Du Pont and W. R. Grace (which had bought Agracetus on the strength of the gene-gun technology) signed a cross-licensing agreement. Acknowledging Du Pont's prior claim to the technology, W. R. Grace ultimately agreed to license its gene gun. In exchange, Du Pont won access to the improvements made by the Agracetus team.

"The Broadest Patent"

In a patent system that was born of the cotton gin and came of age with the light bulb, one might think that proprietary ownership rights to the invention of a working machine might represent the big prize. But this is emphatically not the case in a world where the broad means of controlling knowledge assets is proving far more lucrative than a new machine or tool. While the W. R. Grace–Du Pont agreement resolved the ownership of the gene gun itself, jockeying had only begun for rights to the intellectual frontier the technology opened up.

As a coalition of public-interest groups would eventually state in a formal complaint to the European patent office, "Using the particle acceleration method for genetic transformation of major food and industrial crops, Agracetus has managed to convince U.S. and European patent examiners that the use of any other genetic transformation technique used on any other germplasm of the same species would amount to a violation of the company's patent rights." In other words, while it may have settled with Du Pont on who owns the gene gun, Agracetus is seeking a far bigger prize: control over the *idea* of genetically manipulating soybean varieties based on the success of their earliest gene-gun experiments.

Agracetus's Smestad argues that when the company began in 1981, "no one knew how to get a gene into a plant." His company's researchers "invested a lot of time and energy coming up with a tool kit" that could be broadly applied to making many improvements to a crop. Smestad's broader vision would place Agracetus in a position of heretofore unprecedented power. As he outlined to the trade press, Smestad foresaw a system that amounted to a kind of contract farming: his firm would hire farmers, who would grow the crops and receive a share of the earnings from Agracetus's genetically improved seeds.

However, with only a tiny percentage of the population in the industrialized world involved in farming and many seed varieties already privately owned, neither the award of sweeping ownership rights over a major crop nor Smestad's grandiose plans for contract farming made more than a ripple outside trade circles. Thankfully for the rest of us, however, the Agracetus soybean patent is exactly what the Rural Advancement Fund International is in business to monitor. Hope Shand's tiny office of four staff members in Durham, North Carolina, far from the centers of commerce or political power, assesses the impact such developments will have upon people who make their living actually growing crops.

Shand got word of the pending Agracetus soybean patent in Europe from a well-placed insider shortly before its announcement. She wasted no time moving into action. "There are many battles to be fought in agriculture today," she says, "but when our organization got word of the Agracetus soybean patent, we didn't have an ounce of hesitation. We knew we had to challenge it."

As Shand was well aware, the soybean patent was not an anomaly. A battle was already under way in the United States over similar patent claims by Agracetus to transgenic cotton. As in the soybean case, the company contended that its pioneering work in using the gene gun to engineer cotton crops entitled it to broad ownership rights over any genetically engineered cotton. In its passage through the bureaucracy, the cotton patent mirrored the soybean patent: Agracetus's soybean claims, which had won a patent in Europe, are still pending in the United States. Conversely, the company's patent on transgenic cotton, still pending in Europe, stunned agriculture circles by winning a U.S. patent at the end of 1992. Caught off guard, Shand's organization mounted opposition after the fact against what she terms "the broadest patent ever granted in the United States in the field of plant biotechnology."

She even got some help from Agracetus: soon after winning the cotton patent, the company drew many reluctant participants into the debate by stating that it would not license the patent for most areas of cotton research and development. In other words, it would not allow any potential competitor to genetically engineer new qualities into cotton. Dozens of labs faced the prospect of having to discontinue their cotton experiments entirely, including a good number conducted under the auspices of the U.S. Department of

Agriculture (USDA). If the patent held up, asserted Melvin Oliver, a plant pathologist at the Cotton Systems Research Laboratory in Lubbock, Texas, it would "shut our work down, hinder research into transgenic cotton, and hurt U.S. agriculture."

Within months the USDA took the unprecedented step of legally contesting the patent, arguing that its researchers, including some at the Lubbock laboratory, had made well-known advances that contributed to the Agracetus technique. According to Howard Silverstein, the USDA's deputy assistant general counsel for patents, the department worried about the breadth of the Agracetus patent and its "adverse impact on developing new types of genetically altered cotton varieties." The incident marked the first time the Agriculture Department had ever joined the fray in a patent controversy.

The history of Agracetus's cotton patent, Shand says, informed her organization's pending challenge to the European patent office over the soybean case. "We question the very foundation of what is going on here," she says. "Quite frankly, I don't believe anyone should be able to own rights that afford such power over an important crop. Equally important, though, we question how the single series of experiments Agracetus conducted using one technique could possibly justify such broad ownership claims. It seems increasingly clear in cases like these that something is terribly wrong."

Shand is not alone in this perspective. Writing shortly after the cotton patent was awarded, Karol Wrage, editor of the trade journal *AgBiotechnology News,* noted, "If most important cotton varieties in the future are genetically engineered, then does this mean that Agracetus will have, more or less, 'patented cotton,' just like Polaroid has patented its instant processing cameras?" The analogy is pertinent because Polaroid not only won a broad patent over the concept of instant photography but, in one of the largest and most expensive patent infringement cases ever, successfully enforced its ownership rights. Its competitor, Eastman Kodak Company, was forced to pay a $1 billion settlement after it had begun to market instant photography cameras in 1976, even though Kodak's models were engineered significantly differently.

Could a similar fate befall some of the world's major crops? As Shand explains it, Agracetus's patents would have profound consequences: "Depending on the licensing and royalty fees involved, these patents could effectively put transgenic research on cotton and soybeans out of reach for many companies and public researchers who are outside of the big league players." And, she warns, with a broad portfolio of patents still pending, the company's monopoly control over transgenic crops may well soon extend to rice, corn, and other critical food crops.

Owning the Code

How is it that any party can own things like as yet unspecified varieties of cotton and soybeans. Sociologist Jack Kloppenburg stresses the importance of a previous generation's agricultural breakthrough: hybrid crops. By developing high-yield hybrid corn varieties in the 1920s and 1930s, plant breeders changed the way growers perceived the seeds they planted. Hybrid corn gave farmers yields as much as 50 percent higher than regular corn varieties. But because hybrids do not yield the same results in successive generations, for the first time farmers were obliged to buy seed from the seed companies each year.

But as great a boon as hybrid corn proved to be for the seed companies, self-pollinating crops, such as soybeans, wheat, cotton, and oats, were not so easily manipulated. Changes bred into these types of crops would extend to all successive generations. As Kloppenburg chronicles, plant breeders soon recognized that the only way to replicate the profitable arrangement they had secured with hybrid corn was by bolstering a legal framework that could protect their crop improvements. Only then, Kloppenburg writes, could the seed industry "realize one of its longest-held and most cherished goals: to bring *all* farmers in *all* crops into the seed market every year."

Congress first allowed patent protection for specific plant varieties in 1930, and since then breeders have sought to steadily enlarge the legal framework that protects them. A succession of court decisions and laws has made it possible to patent life forms, especially plant varieties. But a key milestone in this legal history is a bellwether 1980 case called *Diamond v. Chakrabarty*, which is to plant breeders what *Brown v. Board of Education* is to civil rights activists. In this case, the Supreme Court ruled, by a five-to-four majority, that a live, genetically altered microorganism could be patented. The landmark decision held that the issue of whether an invention is animate or inanimate has no bearing on its patentability as long as it meets the criteria of being a novel, useful product of human manufacture and not obvious to someone skilled in the particular field in which the patent is filed.

Ananda M. Chakrabarty, a microbiologist then at the General Electric Company, had used genetic engineering techniques to create a novel bacterium that presumably had never before existed in nature. Because Chakrabarty's bacterium showed promise in breaking down crude oil, its lucrative potential for cleaning up oil spills or remediating toxic waste led him to file for what is known as a utility patent in 1972. While specific plant varieties had long been protected through a more limited category of patent, Chakrabarty's claim, relying on the power of genetic engineering — as yet untested in the courts — asserted that the bacterium should be treated much like any useful machine or invention. As Chakrabarty's legal team argued, he had invented something that could produce results for its users.

The U.S. Patent Office initially rejected Chakrabarty's application, ruling that bacteria, as living organisms, were not eligible for a utility patent. The agency contended Chakrabarty's bacterium was "a product of nature"; allowing such a patent, the examiners argued, would stretch the Patent Office into areas it was never intended to regulate. But Chakrabarty's quest for ownership prevailed eight years later, when the Supreme Court ruled that the bacterium did in fact meet the statutory definition of patentable subject matter. Harking back to the founding principles of the U.S. Patent Office, the high court ruled that "anything under the sun that is made by man can be patented."

Not surprisingly, debate over the *Chakrabarty* decision has continued in many circles ever since the ruling was handed down. Some have questioned the notion of patenting life forms in the first place, arguing that because life represents a continuum from bacteria to mammals, the logical extension of the *Chakrabarty* decision is that even humans might eventually be legitimately patented. Others, like Mark Sagoff, a philosopher at the University of Maryland, have criticized the patent by emphasizing the importance of the role nature plays in genetic engineering research. In a 1996 essay Sagoff argues that Chakrabarty did no more than manipulate nature's design. The altered organism, Sagoff says, may be useful, but "the element of design resides in nature."

Lofty considerations aside, some thoughtful attorneys have also challenged the merits of the *Chakrabarty* decision on its particulars. John Barton, for instance, director of the Law and High Technology Program at Stanford Law School and a legal scholar specializing in patents, believes that the Supreme Court's ruling set the stage for legal decisions in the biological sciences that draw inappropriate precedents from patents granted in the pharmaceutical and chemical industries. Barton contends that genetically engineered products differ from the chemical compounds patented by inventors in the past. In biotechnology, he argues, "analogies are so easy to draw between one advance and another" that it is often fundamentally unclear "how far to go in defining the scope of a patent." He notes, for example, that the courts have yet to address adequately the fact that two "genuinely distinct" genetically engineered organisms may produce the identical protein. Plus, Barton says, he worries that monopolistic claims, such as the European patent granted to Agracetus on transgenic soybeans, run counter to the patent system's goal of "encouraging innovation" and will squelch important areas of research in genetic engineering.

Despite these various reservations, the juggernaut of expanding intellectual property rights over the conceptual realm continues. And, given a court system based on legal precedent, there is little opportunity to reconsider the expanding nature of patent claims in the biological realm except in the legislative sphere. In the years since *Chakrabarty,* the courts have considerably strengthened and expanded the rights of inventors to own species of living organisms. Most notably, such ownership claims were further solidified in 1987 when two Harvard University researchers, Philip Leder and Timothy Stewart, won a patent on a "transgenic mouse," a species that had never before existed. Leder and Stewart "created" the mouse species by inserting a cancer gene into mouse egg cells.

Critics of the idea of owning higher life forms see it as more than coincidence that the first privately owned species is a doomed race, a strain of mice that, as Leder describes it, "has taken the first step toward cancer." But, as with *Chakrabarty,* the so-called Harvard mouse represented another seemingly logical step: the strain of mice was novel, useful, and had not yet been undertaken by anyone else. In granting a patent to Leder and Stewart over the entire disease-prone strain of mice, the Patent Office merely took its cues from Congress and the Supreme Court. Leder and Stewart's mice, as well as successive generations of their ailing offspring will all be the private property of this biological team for nearly two decades.

Today the outcomes of the patent disputes over soybeans and cotton are still unclear. In the soybean matter, the European patent is technically in effect pending resolution of a court case contesting it, while the U.S. Patent Office rescinded the cotton patent after a reexamination process, setting in motion a lengthy appeals process. Legally, research on genetically engineered cotton and soybeans can continue in the United States today, but the specter of broad private ownership claims remains. Both cases will likely drag on for years to come.

Because the precedents involved in these cases are so important, the absence of a strong countervailing power to advocate on behalf of the public interest is glaring. The Rural Advancement Fund International has certainly tried to champion this cause, but as Shand says, "We're tiny almost to the point of being insignificant on a world scale and we're often swimming against a strong tide." Meanwhile, as the giant corporate titleholders face off against one another, the patent cases can become not only byzantine but distorted, as can be seen clearly in the soybean case.

In her battle against the Agracetus soybean patent, Shand was joined by a number of the company's competitors, most notably the large agribusiness firm Monsanto. Although the soybean patent itself received little coverage outside of trade circles, the odd alliance between Monsanto and Shand's tiny organization did draw the attention of the *New York Times,* which noted that the patent was drawing fire from two parties that disagreed on almost every conceivable agricultural policy issue. It was, of course, the shakiest of alliances, prompted by a common enemy.

Exercising her rights under the European system, Shand recruited sympathetic lawyers to help draft a complaint on behalf of the Rural Advancement Fund International, which, along with Monsanto and a handful of other interested parties, contested the patent, throwing the matter into court. Today, however, as Shand readies her arguments in the case, her organization is more alone than when it set out. One-time ally Monsanto took a different tack in the battle, adopting the time-honored "if you can't beat them, join them" strategy: in a $150 million deal with W. R. Grace in April 1996, it simply acquired Agracetus's "plant biotechnology assets and related intellectual property." Monsanto, which, after all, cannot be expected to challenge itself in court, quietly withdrew the 292-page statement of opposition it had filed at the European patent office. "We called them as soon as it was announced that Monsanto had bought Agracetus's plant patent portfolio," Shand recounts with a chuckle. "The public affairs people hadn't been briefed yet about how to handle the company's reversal of its position. They simply didn't know what to say."

In the 1994 press release announcing its opposition to the Agracetus patent,

Monsanto had argued that the patent should be "revoked in its entirety" because the genetic engineering techniques employed by Agracetus were well known in the field at the time. According to Monsanto, Agracetus researchers had not made the novel "inventive step" needed to justify such a broad claim over transgenic soybeans. As Monsanto spokeswoman Karen K. Marshall explained, "The claims are simply too broad given what was going on in genetic engineering at the time they filed for the patent." Today Marshall offers only uncomfortable silence on the "simply too broad" question. Her one sanctioned comment about the contested European soybean patent is: "Monsanto will defend it."

License to Sow

Legal fights over knowledge assets in agriculture may sputter, stall, and delay en route to the creation of empires of intellectual property, but the imperatives of nature and the press of modern food production cannot be so delayed. Far removed from the courtrooms and boardrooms where ownership claims are deliberated, farmers still plant soybeans as they always have. But today, as John McClendon can attest, life for the farmer has changed.

McClendon is an Arkansas soybean farmer with 4,500 acres and forty years of experience, including a stint as the CEO of an insurance holding company. An independent thinker, he is a good example of his own adage that "if you've met one farmer, you've met one." McClendon raised some eyebrows in 1996, the year after his term as president of the American Soybean Association, when he refused to sign a contract Monsanto offered to farmers wishing to purchase the company's genetically altered Roundup Ready soybean seeds, which have been altered to tolerate higher doses of the company's best-selling herbicide Roundup. Monsanto's motivation is clear: they can charge a premium for the seeds and sell more herbicide at the same time. The farmers' motivation to buy such seeds is equally straightforward. As McClendon explains, on his farm he has largely been unable to control two particular species of weeds: "There's just nothing there to handle these two," he says. If Monsanto's genetically engineered seeds allow farmers to apply enough herbicide to control the weeds without killing the crop, the additional soybean yields could easily offset the added costs.

It was not the price of the seed or even the five-dollar-per-bag "technology fee" that Monsanto charges for its genetically altered soybeans that upset McClendon. It was the provisions of the agreement the company makes farmers sign. As he puts it, at the first moment he read it, the Monsanto contract struck him as "outrageous."

"I've got a strong background in contracts, and this one didn't strike me as very good from my perspective. From the beginning I expressed my opinion to higher management and I never signed it." What did Monsanto require that would alienate a stalwart, corporate-friendly farmer like McClendon?

Monsanto's 1996 Roundup Ready Gene Agreement is a license contract between the farmer and the company. To get the high-tech seeds, the farmer must relinquish his or her right to save or replant the patented seed or to sell seed derived from it to anyone. Using the seed for any purpose other than growing the crop is strictly forbidden. According to the agreement, a farmer who does reuse the soybean for seed faces penalties equal to "100 times the then applicable fee for the Roundup Ready gene, times the number of units of transferred seed, plus reasonable attorney's fees and expenses." In addition, the agreement adds, Monsanto can seek unlimited damages for any violation. A farmer who violates the contract could lose his or her farm and all other assets — a steep penalty to pay for saving or replanting patented seeds.

The provision that irked McClendon most, however, was the right Monsanto claimed to monitor a farmer's land. Specifically, the 1996 agreement states: "Grower grants Monsanto, or its authorized agent, the right to inspect and test all of Grower's fields planted with soybeans and to monitor Grower's soybean fields for the following three years for compliance with the terms of the Agreement." As McClendon complains, "However good the seeds might be, I didn't want Monsanto representatives feeling like they could come on my farm any time they wanted and dictate to me what to do on my land."

In 1997, however, after negotiating some changes in the contract, excising the provision allowing company representatives unlimited access to his farm, McClendon has decided he will plant some Roundup Ready seeds. He is willing to bet on the seeds even though yields of Monsanto's related Roundup Ready cotton have proved a dismal failure for many farmers in the southeastern United States. "They spent a lot of money developing these seeds," McClendon reasons, "and technology is what will keep U.S. agriculture second to none."

McClendon says he worries about a lot of things that are happening in agriculture today. He worries about consolidation in the farming industry: a nation that once boasted some 2 million farmers now makes do with 600,000. Some 20 percent of the farmers grow some 80 percent of the food in the United States. He worries that we are returning to "a plantation era," in which farms have to keep growing larger to meet economic demands. "I've got 4,500 acres now," he says, "but I imagine I'll need 10,000 to stay in business five years from now."

But McClendon says he can't afford to put such broad considerations — or the warnings he's heard from people like Hope Shand about an eventual return to sharecropping or "bioserfdom" — above the need to keep his farm economically viable. "No matter what all's going on," he says, "I've still got to figure out how I'm going to plant next year's crop."

But while farmers like McClendon can't be expected to turn back a strong tide alone, the rest of us need to understand the wide-ranging effects of the changes in agriculture over the past decade. The genome of food crops like corn or soybeans is a common heritage of humanity that quite literally nourishes and sustains us. In an unprecedented manner and with alarming speed, we have traded away a time-honored, decentralized, diversified form of agricultural husbandry that allowed farmers to help steward crop varieties and guide their improvement. In its place we have allowed a highly centralized, multinational system driven by the pursuit of near-term profit to become dominant.

Sadly, some of the results of these developments are clear already in the type of seeds Monsanto is licensing McClendon to sow. Roundup Ready soybeans — whether they ultimately succeed as a crop or not — are designed to tolerate levels of Monsanto's Roundup herbicide that would kill other varieties. Clearly, the seed variety offers a double boon to Monsanto, but it is dubious that it advances our global agricultural interests of sustainable farming practices that reduce reliance on costly and environmentally hazardous inputs.

We have acquiesced in these changes largely in response to the relentless quest for innovative crops and higher yields. But the root of the problem resides in our inability so far to set commonsense boundaries on what conceptual tools can be privately held. Instead, we have blithely allowed companies to stake broad claims over the entire genome of important global crops.

Ironically, the change is likely to lead to higher food prices because of the monopoly position held by many of the firms in the emerging global seed cartel and because of the fees farmers and companies alike will have to pay to license the intellectual property needed to create crops with sought-after traits. The change also portends a threat to our "food security," as control over new varieties moves to the proprietary grasp of a shrinking number of firms. As many agricultural scientists have long pointed out, this kind of consolidation and widespread reliance on a limited number of engineered crop varieties heightens our susceptibility to blights, rusts, and other crop diseases that regularly emerge and greatly increases the potential damage they can cause.

But the biggest change of all is the insertion of a powerful new wedge of private ownership over something formerly shared. And the effect for all of us, farmers and nonfarmers alike, is to needlessly and even dangerously separate us further from the land and the growing of food that sustains us all.

7
Taborsky's Lament

JANUARY 1996, TAMPA, FLORIDA. It was inconceivable to Petr Taborsky, throughout his long ordeal, that he would wind up in prison. But now, in the bright winter glare of a Florida afternoon, that is precisely where he found himself: shackled at the ankles amid a pack of men, hacking at thick brush and stumps with a hoe to clear debris under the stern gaze of shotgun-toting prison guards. Even here, on the chain gang of a maximum-security unit of the North Florida State Penitentiary, the swirl of events that had landed him in jail seemed nearly beyond belief. Taborsky was incarcerated among drug dealers, robbers, and sex offenders, but his crime was decidedly different: he had been found guilty of stealing his own research.

Taborsky is the first person in U.S. history to be jailed on criminal charges for stealing an idea unrelated to state secrets or espionage. His tragic tale is encapsulated in the two glaringly different numbers the government has issued him. First, there is U.S. Patent No. 5,082,813: the first of three patents awarded to a bright young undergraduate at the University of South Florida for inventing a reusable cleanser that can remove ammonia from wastewater. Then there is his prisoner number, 514,527, issued by the penitentiary where Taborsky was incarcerated. The numbers betray a deep confusion and ambivalence in our society about a researcher's right to his or her ideas as we move toward the twenty-first century.

"It's remarkable," says David Lange, a professor of intellectual property at Duke University, of the Taborsky case. "It's hard to imagine circumstances sufficiently exacerbated under which a university would sue one of its students, much less pursue criminal charges."

In the most immediate sense, Taborsky's tale offers a caricature of the American system of corporate-sponsored university research and intellectual property rights. Although the circumstances of the case are exceptional, it underscores what can happen when universities, beholden to industry for an increasing share of research dollars, let financial concerns overshadow the notion of research as a shared intellectual pursuit.

More than that, though, Taborsky's case can be seen as a powerful allegory about the fragility of the remaining realms of shared knowledge in the emerging economy of ideas. Our universities, libraries, and even museums, bastions of the beleaguered "conceptual commons," find their administrations sorely tested in the face of private commercial pressures. In this sense Taborsky's case — extreme as it is — merits our close attention to understand what is at stake in this new climate for all of us.

A Researcher's Conviction

As far as the U.S. justice system is concerned, Taborsky's tale begins in 1987 when, as an undergraduate at the Tampa-based University of South Florida, he took a job earning $8.50 per hour as a research assistant in an engineering laboratory managed by Robert Carnahan. Carnahan, the dean for research at the university's College of Engineering, whose long career had already earned him many publications and several patents, had contracted to conduct research paid for by a local holding company called Florida Progress Corporation.

Carnahan's arrangement, known as "sponsored research," is increasingly common at universities around the country. As distinguished from more general "no-strings" grants, which an academic department or research laboratory can use as it sees fit, sponsored research projects tend to follow a fairly standard arrangement between the university and the firm or sponsoring agency. The project Taborsky was involved in was no exception. Under the terms of the agreement, if inventions or patents resulted from the research, the university would retain the patent, but the firm would retain a preferential right to license the technology.

The research in this case could hardly be considered the standard fare of high finance or high technology. Florida Progress had paid the university the modest sum of $20,000 to conduct research on a claylike substance called clinoptilolite, commonly used in cat litter. The company was interested in the material because it is inexpensive and can absorb ammonia, making it potentially useful in filtering chemicals from wastewater in sewage treatment plants. Clinoptilolite, however, becomes saturated relatively quickly. By funding the university research, Florida Progress hoped to find a way to increase the amount of ammonia the clay could absorb.

By all accounts, the company was lucky to have someone as talented as Taborsky addressing the problem. A stiff and awkward young man, Taborsky had shown promise in science at least since age six, in 1968, when his family fled the Russian invasion of Czechoslovakia to come to the United States. In high school he scored in the ninety-eighth percentile on a national placement test in organic chemistry and was the class valedictorian. At the University of South Florida he studied both chemistry and biology in the hopes of eventually becoming a neurologist.

Ultimately, beginning in 1992, as a result of the research problem Florida Progress had brought to his attention, Taborsky was awarded three U.S. patents. He remains proud of the accomplishment, but the patents have caused him nothing but woe. "My life has been made miserable at the hands of my university and a corporation," Taborsky says. "I've been humiliated. My credibility has been damaged."

The hotly contested versions of what happened in this case diverge in the latter half of 1988. At the end of the three-month period of sponsored research on behalf of Florida Progress, Taborsky contends he received Carnahan's permission to pursue his own experiments. Florida Progress had sought some type of bacteria or enzyme that could be added to the clinoptilolite to make it more absorbent. For his own research, however, the twenty-six-year-old Taborsky decided to try a different approach; he spent the summer trying to understand clinoptilolite's chemical properties. A semester short of graduation, he hoped he might use the research as the basis for a master's thesis. He says he discussed the project with Carnahan.

Spending many nights in the lab, often waking every twenty minutes to monitor his experiments, Taborsky had a breakthrough that summer. He found that clinoptilolite was able to absorb more ammonia after it was heated to temperatures above 800 degrees Fahrenheit. Previously no one had been able to keep the material from breaking down completely at such hot temperatures to observe the effect.

Taborsky knew immediately that his results were both important and potentially lucrative. The question was: who had a legitimate right to the invention? The firm claimed that the discovery grew out of its sponsored research; Taborsky's professor told him the invention belonged to the university, which had "shop rights" to any work undertaken in a campus lab. Taborsky, however, believed intensely that his work, undertaken largely on his own during a summer work-study project, was a novel departure that owed little debt to anyone. As Taborsky recalls, "I decided I wasn't going to let them intimidate me."

Initially, when Professor Carnahan reported the discovery to representatives from the university and the sponsoring firm, a subsidiary of Florida Progress offered Taborsky a job and told him he could be the primary author in their patent application. Taborsky was tempted enough by the deal to sign a routine confidentiality agreement with the company — the first such agreement he had signed. But the arrangement quickly fell through when, after consulting a lawyer, Taborsky realized the contract offered him no real guarantee of employment. The agreement, he says, stated that the company could fire him at will. At this point, in the fall of 1988, he alarmed all parties involved by threatening to seek his own patent on the clinoptilolite discovery.

Taborsky became increasingly concerned that he would be cheated out of the reward for his hard-earned research results. In December he left school and his lab job, missing his final exams. Outraged that an underling in a sponsored research project would threaten to stake a private ownership claim on his research, university officials pressed criminal charges of theft against him, and within weeks he was arrested. According to a police report, Carnahan claimed that Taborsky had taken two research notebooks from the university laboratory in violation of a confidentiality agreement. Taborsky admits that he took the notebooks. But he contends that they belonged to him. He had personally purchased them from the nearby campus bookstore, he says, and he took them home with him almost every night when he was working on a separate strain of independent research. And he rightly argues that he never signed any agreement with the university.

The case soon became desperately polarized. Most of those who are familiar with the details say they remain baffled that the parties couldn't reach some kind of agreement. Taborsky says that when university officials initially pressed criminal charges against him, he felt that they were simply trying to intimidate him — a tactic, he says, that only heightened his resolve not to give in to the administration. But Taborsky paid a high price for his obstinacy. His U.S. citizenship application was put on hold, his marriage crumbled, and the university even withheld his chemistry degree.

Before the saga was through, these legal proceedings landed Taborsky in jail. Convicted by a jury of stealing university property, he was given a three-and-a-half-year sentence. As part of Florida's new penal system, his felony conviction required him to begin serving his sentence on a chain gang, starting in January 1996. But in response to news stories depicting a promising young scientist in shackles, Florida's governor, Lawton Chiles, stepped in. As Dexter Douglass, legal counsel to the governor's office, told the press: "There are a lot of things in this case that raise your eyebrows. We are concerned that the government overreached in this young man's case." After reviewing the case, Governor Chiles even offered Taborsky clemency, but he refused, seeking instead to have what he called his "wrongful conviction" overturned. Thanks to the governor's intervention, however, Taborsky was transferred from the chain gang to serve out the remainder of his sentence at a minimum security prison. And with time off for good behavior, he was released in the spring of 1997 to face eleven years of probation.

Since Taborsky's release, he has pursued a tangle of civil legal actions stemming from his criminal case, motivated by his deep desire to redeem his scientific integrity and personal honor. The university, having already reportedly spent more than ten times the amount of Florida Progress's grant on outside legal counsel alone, is suing Taborsky over the remaining two patents, claiming ownership. He, in turn, has sued the university, its board of regents, and two principal investigators personally, including Carnahan.

Even if one follows the inexorable and myopic legal proceedings, it is hard to imagine how a fight over an idea for making cat litter more absorbent could ever have resulted in a chain gang, a governor's intervention, or a succession of bitter, costly lawsuits. Few close observers of the Taborsky case would fully absolve any of the parties from responsibility for the tragic outcome so far. But perhaps most alarming is the aggressive stance pursued throughout by the university. Consider, for instance, the actions of Francis Borkowski, president of the University of South Florida, during the initial rounds of the legal case. At a key juncture Borkowski, presumably frustrated by Taborsky's recalcitrance, appealed personally to the judge to sentence him to jail. Why would a university president intervene in this way? Why would the university go to such lengths — and expense — to prosecute a case involving an undergraduate and a fracas over a research grant so small that it would barely cover annual tuition and expenses for a single student?

Henry Lavendera, an in-house attorney for the University of South Florida, candidly explains that the university feared the case might scare off future research sponsors. As Lavendera puts it, "We are concerned that potential sponsors will view it as a black eye for the institution if we allow student researchers to steal information." According to Lavendera, the case is no different from that of a student stealing precious books from the university library. "It is not my characterization but a fact that Taborsky is a convicted felon," Lavendera says. "The university has taken heat in the media that Taborsky wound up in prison, but the fact is he stole property that didn't belong to him."

Sadly absent in the remarks of all the parties involved is any mention of what is traditionally understood as the primary mission of a university: to foster a learning environment in which people collaborate in the research process.

At the University of South Florida, the second largest state university in Florida, sponsored research has more than quadrupled in the past decade or so: from $22 million in 1986 to roughly $100 million today. This explosive growth is mirrored at universities across the country. Overall, corporate-sponsored research on university campuses has more than doubled since 1988, now accounting for roughly $1.5 billion per year, according to the National Science Foundation.

"This case is just the tip of the iceberg," says Jerome Reichman, a law professor at Vanderbilt University familiar with the Taborsky affair. "And it is going to get much worse." As Reichman explains, universities have a serious problem. They risk a brain drain if their academics are not suitably remunerated for their creative inventions, and they're starving for basic research. Disputes such as Taborsky's (minus the criminal charges) are increasingly acrimonious and even vicious, say Reichman and other experts, particularly at universities where money from corporations has replaced traditional sources of funding. "In the old days, it was almost beneath a tenured professor to file for a patent," Reichman says. How times change. Even though the federal government still funds the majority of research projects at universities in the United States, intense interest in patenting has now become the norm, as companies are guaranteed protection, exclusive access, and rights to new technologies or medical breakthroughs.

Arrested Inquiry

Cornelius Pings, president of the Association of American Universities, a Washington-based nonprofit consortium of the nation's largest research universities, notes that despite what he calls the "bizarre" nature of the Taborsky case, the underlying tensions that it exposes are real and mounting. Today's universities, Pings explains, can no longer "maintain they are above commerce." And yet, he says, they are expected to — and must — protect their role "as temples of intellectual inquiry."

It is, to say the least, a tall order. Long viewed as fertile havens for nurturing and cross-pollinating ideas that would be profitable only in future generations, the nation's research universities have come to be seen by corporate sponsors as tempting orchards of unclaimed conceptual fruit for the picking. This is a sorry state of affairs, not because it facilitates the transfer of knowledge from the ivory tower to the marketplace — which in and of itself is a good idea — but rather because it gravely erodes the conceptual commons and impoverishes the university's fundamental educational mission.

The situation is all the more striking because the research universities of the United States have long been the envy of the world. As Charles Vest, president of MIT, noted in a recent speech to the National Press Club, they "are the foundation of our entire national research infrastructure." As Vest and others have long noted, university research can be credited with a great many developments, from polio vaccines and cancer therapies to jet airliners and computers. Our broad-minded federal support for university research, Vest rightly observes, "is an investment in the future of our human capital — people and their ideas."

Lest there be any doubt about it, a growing body of quantitative data confirm the importance of universities in fueling economic growth. For instance, one of the first longitudinal studies of the economic impact of a research university, conducted in 1997, found that over the past several generations, MIT graduates have founded 4,000 firms, which in 1994 alone employed 1.1 million people and generated $232 billion in sales worldwide. If the companies founded by MIT graduates and faculty formed an independent nation, according to the report, their combined revenues would rank it as the twenty-fourth largest economy in the world.

Now, however, as the Taborsky case indicates, this winning formula for innovation is being jeopardized by proliferating and often oppressive private claims over the useful ideas and concepts incubating at universities.

At universities across the country, a steady stream of intellectual property disputes has begun to emerge. In Providence, Rhode Island, in 1997, an occupational health physician at a Brown University–affiliated teaching hospital claimed that the hospital tried to suppress his research on an outbreak of lung disease at a local textile plant. The researcher, David Kern, claims that when he published his findings, the hospital closed his clinic in retaliation. Kern initially discovered the prevalence of lung disease when he was under contract with the textile firm Microfibres, Inc., of Pawtucket. But he says he soon discovered that his research results were unwelcome. The company threatened to sue him for seeking to publish the data. Company lawyers argued that Kern was breaking a secrecy agreement, even though Kern's paper did not identify the company by name or include any proprietary information.

In Texas, a battle arose in 1995 over software that researchers believed could compress video images more than a hundredfold. The exciting commercial possibilities included the potential to fit feature-length movies on a single CD-ROM disk and allow video teleconferencing to take place over regular phone lines. As it turned out, the researchers' claims for the software were overblown. But that didn't prevent a bitter three-way legal battle among the young computer scientist who actually wrote the code on his laptop computer, the mathematician at Texas A&M University who laid the conceptual foundation for the software, and the quasi-private institution called the Houston Advanced Research Center, which underwrote much of its development. At one point police were called in to confiscate the programmer's laptop computer, believed to contain the only known version of the newly minted software.

In Seattle, concern about these kinds of commercial disputes prompted a group of scientists at the University of Washington to formally call for stronger protection for scientists who make discoveries that threaten vested interests. Writing in the *New England Journal of Medicine,* the researchers relayed personal experiences of being attacked by funders, advocacy groups, and other vested interests for unwelcome research findings. Despite the abundance of anecdotal evidence, surprisingly little quantitative research has yet been undertaken to document the problem. But the existing data largely substantiate the growing concerns of many researchers.

One 1997 study examined companies involved in research in the so-called life sciences — such as pharmaceutical, agricultural, and chemical firms — and found that 90 percent of the 210 companies surveyed had financial ties to university researchers. For 60 percent of these companies, the relationships yielded patents or technology applicable to new products for the companies. Notably, a full third of the companies reported having disputes with their academic partners over intellectual property. Twelve percent even said they believed their academic research partners may have engaged in professional misconduct or questionable scientific practices.

Another set of researchers reviewed the publications of more than 1,000 scientists at universities in Massachusetts. This 1997 study found that slightly more than a third of these articles had one or more authors who stood to make money from the results they were reporting — either by holding a key patent or by serving as an officer or board member of the company exploiting the research. Yet none of the papers mentioned that the authors had a financial interest in the results.

The increasingly commercial orientation of research at the nation's universities marks a fundamental shift toward private exploitation of concepts and know-how. This change, especially in the past twenty years, stems from the 1980 passage of the Bayh-Dole Act, the landmark bill that gave educational institutions ownership of inventions created with federal funds by their faculty and research affiliates.

At the time, preoccupied with competition from high-tech Japanese industry and fears over a sluggish domestic economy, U.S. legislators sought to more quickly transfer the fruits of university research to the marketplace. In essence, the universities were freed to launch for-profit entities to facilitate the transfer of technological developments to the marketplace. Many opponents of the bill worried that it was a sellout to corporate interests and would have the undesired consequences of undermining academic independence and scientific integrity. In the aftermath of passage of the act, then-Congressmember Al Gore fretted publicly that the arrangement risked "selling the tree of knowledge to Wall Street."

Among the critics of the Bayh-Dole Act was Admiral Hyman G. Rickover, who testified that he believed it represented one of the greatest giveaways in American history. "Based on forty years' experience in technology and in dealing with various segments of American industry, " Rickover warned, "I believe the bill would achieve exactly the opposite of what it purports," hurting small businesses and stifling competition while promoting "greater concentration of economic power in the hands of large corporations."

Today, nearly two decades later, the Bayh-Dole Act is widely touted as a success story, rescuing publicly funded research from languishing in the ivory tower. But Rickover's concerns are echoing ever louder in some quarters, in light of the pressures the act has brought upon open, unfettered research.

For one thing, the vast majority of university researchers decry the creeping scourge of secrecy that forestalls the rapid diffusion of new ideas and developments in a field. One research group at Massachusetts General Hospital found secrecy to be particularly widespread among biomedical researchers. Out of some 2,000 academic life scientists surveyed nationwide in 1997, 79 percent said they had delayed sharing new information in order to apply for patents or secure some other kind of intellectual property protection for their work. A fifth of those surveyed reported that for commercial reasons, they had personally delayed publication of their data for more than six months, in some cases keeping it secret indefinitely. Even more troubling, nearly a third of those who delayed publishing research results admitted they had sought to "slow dissemination of undesired results," presumably because of the commercial stakes — such as the effect of the results on stock prices of companies they were related to.

Quantitative studies like these offer an important picture of a changing research environment. But they fail to convey the true costs of secrecy, as played out daily in countless ways. Take, for example, the proliferation in academic biomedical fields of contracts governing the routine exchange of biological materials that serve as the basic building blocks of much biological research. Known as material transfer agreements, or MTAs, these increasingly elaborate contracts between research groups stipulate that in order to gain access to a given biological material — such as a proprietary gene sequence, microorganism, or other genetic resource — the recipients must agree to surrender all property rights on discoveries that might be contingent upon the material's use. Often the agreements also demand the right to preview and comment on any articles that might arise from research involving the material.

MTAs are essentially fences erected between research teams that will only get taller and more common across the research landscape.

The use of MTAs has drawn the ire of many researchers in academia, who liken them to a "spreading virus" of restrictions. Kate Phillips, a staffer at the Council on Governmental Relations, a nonprofit group that represents 142 research universities, calls the increasing complexity of MTA contracts — and the time and effort devoted to negotiating them — "a horrendous problem." Julie Norris, director of the sponsored research office at MIT, complains that "no amount of education" by university researchers and administrators has diminished companies' attempts to gain control or outright ownership of discoveries arising from the use of shared materials. According to Keith Yamamoto, a biologist at the University of California at San Francisco, the situation already "endangers the academic tradition of free and open publication."

All told, encroaching commercial interests pose a serious threat to the open exchange of ideas that is the lifeblood of universities. Some, like Paul Berg, professor of biochemistry at Stanford Medical Center, have gone so far as to claim that commercial incursions are destroying pure, discovery-based science in U.S. universities. As Berg puts it, "We sit here and talk about feeding ideas into American industry, but we ignore the price we will pay."

One confounding feature is the difficulty of assessing the value or worth of noncommercial university research when ideas become seen as commodities to exploit. As an editorial in the science journal *Nature* put it, "Today's fundamental elucidation of a protein interaction can be (almost literally) tomorrow's profit-earning test kit. Today's gene sequence, if not applicable tomorrow, may be assumed to be replete with vast profits achievable the day after." The editorial continued, "Although commerce is to be supported, there are signs of a culture being made sick by commercial interests to an extent that is unprecedented in science." What is needed is a strong and enforceable set of rules governing the ownership of shared conceptual terrain. A key obstacle is the difficulty of sustaining an open environment governing the development and exchange of concepts alongside a commercial orientation toward them. Economist John Kenneth Galbraith has made a similar point: "In an atmosphere of private opulence and public squalor, the private goods have full sway."

In this climate one might predict that universities will go into business for themselves. In fact, such financial arrangements are already widespread. Many major research universities have established for-profit venture capital entities to commercialize the research of their faculty members. For instance, in 1988, Harvard University unveiled a venture capital arrangement called Harvard Medical Science Partners, a for-profit entity to invest in the research of its faculty and fund products for development. The venture has drawn fire even from other university officials involved in similar venture capital funds. Duke Leahey, director of industry contracts and licensing at Washington University, for instance, says that "the basic problem with the Harvard fund is that the institution is involved directly in raising money from investors and now has a fiduciary responsibility to them." Leahey dismisses Harvard's claims that its venture capital arrangement will remain a distinct entity, saying, "We'll see what the lawsuits say later on."

In his final annual report in 1991, Harvard's president, Derek Bok, warned that "it will take very strong leadership to keep the profit motive from gradually eroding the values on which the welfare and reputation of universities ultimately depend." As universities become more entrepreneurial, Bok declared, "they appear less and less as charitable institutions seeking the truth and serving students and more and more as huge commercial operations that differ from corporations only because there are no shareholders and no dividends."

Concerns like Bok's have prompted some well-known observers — including Sir John Maddox, former editor of *Nature* — to publicly question whether the research university can even survive in such a commercial environment.

Auctioning the Archives

As many have noted, the shift toward private sponsorship of university research represents a response to prevailing political winds: withering governmental research funding and a policy environment that favors privatized solutions to a wide range of social problems. But while many have recognized these underlying causes, too few have focused on the implications of the changes, which extend far beyond the gates of academe. Put bluntly, they represent the pervasive erosion of our ability to freely share ideas. To see another aspect of the trend, pay a visit to your local library. Here, if you look carefully, you will see an institution under siege.

With the growing power of the Internet, and the ability of computer users to pay for access to information on their home and office computers, the idea of a designated place in which information is made available freely to all patrons is now viewed by many as anachronistic. As Bernard Margolis, president of the Boston Public Library, noted in 1998, if you review the evidence, you might even conclude that the library is "washed up, kaput, dead, gone — and if not quite there, at least close enough to call the hospice." Not surprisingly, Margolis believes that the mission of the public library remains vital to our society. The question is, will the free-access mandate continue to be viable in a world where access to knowledge is tightly controlled in order to generate profit?

If openness and ferment have eroded at the nation's universities, a parallel struggle at the nation's libraries underscores a profound rift over the issue of shared information. On the one hand, the advent of on-line computer databases offers a tremendous amount of new information to library users. Yet just as the promise of electronic networks to expand public access to information is becoming a reality, librarians find their options significantly limited by profit-seeking information owners. The broad battle lines have been drawn nationally, but the actual skirmishes take place almost daily at some 16,000 public library branches and another 3,000 academic libraries nationwide.

Researchers at public libraries in Massachusetts, for example, were in for a surprise in the spring of 1998. When they tried to use library computers to access newspaper citations from a vast on-line database called the National Newspaper Index, they found that the service had been terminated. As Alan Allaire, reference librarian at the Newton Public Library, explains, the service had been negotiated through a statewide arrangement providing low-cost service to the state's public libraries. But when it came time to renew the contract, the company had significantly raised its prices. As Allaire explains, "We hate to take information resources away from our patrons, and we're looking for a replacement. But in this case we simply couldn't afford the fees."

Or consider the case of Lloyd Davidson, a librarian at Northwestern University. In the early 1990s Davidson approached a prominent commercial publishing and database firm, seeking to reprint the firm's citations and other bibliographic information about Northwestern faculty members on a campus-wide computer network. After several months of deliberation, company representatives quoted Davidson a price of $30,000 for accessing the data, explaining that since the information would be placed on a computer network and they couldn't be sure exactly how many records Davidson would ultimately make available, they had to charge a fee equal to that for buying the entire database.

"It sounded like extortion," says Davidson, explaining that he merely wanted access to about a thousand short computer records derived from the open literature. When he complained about the price, the firm came back with a new offer: they said he could have the records for free. But they imposed so many qualifications, including a proposed time limit of one year for use of the material, that Davidson abandoned the project altogether.

Library skirmishes like these are commonplace today, emblematic of the contentious broader debate. "Proprietary information and copyright in the computer age run right up against American values of free access to information," says Karen Muller, spokesperson for the American Library Association (ALA). "The issue is cutting to the core of the way librarians have operated for a very long time."

For a variety of reasons, owners of the new electronic media see their role as far different from that of traditional publishers. To protect their rights to the material, more and more electronic publishers are seeking to control access to the information they have compiled, leasing use of the material rather than selling it as a product like a book or magazine. As a result, on-line database owners force libraries into "metered access" arrangements that threaten to change the library from a public repository offering free access to its resources to a mere gateway that charges for information held by others.

The "fee or free" debate that has brewed over the past decade or so revolves around the assignment of costs, either through subsidized payments or direct reimbursements. No one disputes the fact that the most important and extensive databases require enormous investments to compile and maintain. At issue, though, is our commitment to broad, free access to information in the face of arguments such as the claim of the Information Industry Association that "those who make the investments need to be sure that those investments can be recouped, and profit generated for further investment."

Historically, libraries have been protected by provisions of copyright law that say that the rights of a copyright holder are not absolute, that under certain circumstances copyrighted material may be fairly copied and exchanged. So-called "fair-use" provisions, for instance, allow for noncommercial uses of information, such as the distribution of materials by teachers and librarians without payment for every use. By almost all accounts, fair-use provisions are a flexible, commonsense approach that has served everyone well for centuries.

Under this historic legal framework, libraries have been able to obtain copies of virtually any information available and share it freely among their patrons. Library users across the country were assured of free access to a significant chunk of society's storehouse of knowledge — from detailed reference works and collections of data to the latest journals and literature.

Today, however, the owners of electronic information, threatened by the ease of copying that characterizes the digital environment, seek to establish pay-per-view arrangements antithetical to those that have traditionally prevailed at public libraries. As a result, library users lack the assurance of past generations that they can gain access to information. Of course, vastly more information is available today than ever before. But in public libraries, the amount of accessible information as a percentage of the total has already dwindled significantly. Unless we make a public-minded commitment to alter the situation, it is destined to shrink further.

"As much as I hate to see it happen," says Frank Romano, professor of printing management at Rochester Institute of Technology, "I'm sure that by the middle of the next century, libraries will be places people go only to look at old books."

Recently, in fact, the information industry has stepped up its campaign to secure tighter protection for what it owns. Several important court cases have limited the power of database owners to control facts that don't contain a creative component. First, in a case involving the manufacturer of a white-pages telephone directory in Kansas, the court ruled that the firm could not, as it contended, claim exclusive right to publish residents' phone numbers. Similarly, the courts overruled the West Publishing Company's claim of copyright to the compilation of the nation's court verdicts that it publishes.

To many in the industry, like Donald Duncan at the Information Industry Association, these decisions represent a threat to the production of the high-quality database products and services we have come to depend upon. Even worse, Duncan and others claim, is "an international threat from Europe." As Duncan explains, database owners in this country fear that a European Union directive initially slated to take effect in 1998 will usher in a wholesale pilfering of U.S. databases throughout the European market. In fact, they say, given today's technologies and the global nature of the information marketplace, the "misappropriation" of information in proprietary databases in Europe will surely flow back into the United States, undermining the domestic market as well. As Duncan puts it, a "well-established and proven system is now threatened."

But many, like Adam Eisgrau, legislative counsel for the American Library Association, dismiss this contention as an overblown scare tactic to tighten intellectual property protection. "There's a technical legal term for the information industry's argument," Eisgrau says. "It's called crap." As he explains, current copyright law here and abroad affords ample protection for information owners. Clamping down on fair-use provisions, he says, is not going to overcome the almost insurmountable difficulties of enforcing copyright restrictions in cyberspace.

If many experts remain skeptical, however, the Clinton administration has been persuaded that we need stronger protection for compilations of information. The government's position was outlined as part of its plan for the National Information Infrastructure, more commonly known as the information superhighway. According to the administration, "The full potential of the National Information Infrastructure will not be realized if the education, information and entertainment products protected by intellectual property laws are not protected effectively when disseminated via the NII. Creators and other owners of intellectual property rights will not be willing to put their interests at risk."

The Clinton administration's position resulted in a remarkable turn of events at the end of 1996. Sympathetic to the concerns of the Information Industry Association and to those of powerful lobbyists from Hollywood and the popular music industry, all of whom seek tighter protections to discourage the pirating of their property, the administration mounted a campaign to strengthen copyright law that illustrates clearly just how tenuous and vulnerable fair use is in the emerging knowledge-based economy.

Unable to secure federal legislation to expand and tighten copyright restrictions on digital information, the Clinton administration attempted an end run around Congress by proposing changes in copyright law that would have effectively undermined fair use in cyberspace. The administration drafted the changes as proposals for a meeting held under the auspices of the World Intellectual Property Organization (WIPO) in December 1996.

The draft treaty proposed to assign to databases intellectual property rights that go well beyond those in present copyright laws and treaties by making any electronic copying of copyrighted material explicitly illegal — from browsing at an on-line database to sending a newspaper clipping to a colleague. The draft treaty was attempting to protect commercial interests, but it was remarkably insensitive to the implications for scientific research and education. The big losers were those committed to the free exchange of information.

The Clinton administration's action prompted a flurry of criticism. Scores of organizations, from the Association of Research Libraries to the National Science Foundation, sent letters warning that the proposals would impede the flow of information vital to the conduct of scientific and academic research. A hundred law professors from around the nation signed a letter warning of similarly grave consequences. Even the American Meteorological Society got into the act, worried that meteorological data now passed freely in the public domain might infringe the proposed provisions. As the National Research Council explained in a review of the matter, the proposed changes were "antithetical to the principle of the full and open exchange of data," a strong and long-lived principle of the United States government.

Ultimately, the U.S. delegation was forced to abandon its plans for a new treaty. The nation's universities, libraries, and museums, in concert with other organizations looking out for the public interest, showed that they had the potential to wield a powerful influence on matters affecting the exchange of information.

But, Eisgrau says, the fight is not nearly over. As a result of the negotiations in Geneva, the administration has explicitly stated its commitment to protecting the notion of fair use in the digital environment. And yet, in a climate overwhelmingly oriented to a pay-per-view approach, the pressures on free access at the library are sure to continue. As Eisgrau puts it, "We can't take the future of libraries for granted." Given the powerful vested interests that seek tighter private control over digitally transmitted information, he says, "We are absolutely not out of the woods yet."

As of this moment it is unclear what kind of broad, free access will be available at the nation's libraries in the future. Some hopeful signs point to the possibility that more information will become widely available. The National Library of Medicine decided in 1997 to make its huge MedLine database of citations to medical literature available on the Internet for free, in a move billed as a boon to the health of all Americans. Similarly, IBM opted the same year to make its powerful database of U.S. patents publicly available for no charge — a service even the U.S. government does not offer.

The fact is, the nonmarket principle of free and equal access to knowledge is the essence of the library as both a place and an institution. Yet the library is vulnerable to the commercial imperative. As economist Robert Kuttner notes, "There is no good market reason for free public libraries. Many of the people who use them could afford to pay a small fee to check out books." In principle, consumer demand could sustain a commercial system with no alternative, just as people subscribe to cable TV. But the existence of the free public library reflects a basic value in our society: the importance of open access to information and learning.

As the computer hacker and technology critic Clifford Stoll has said, "Visit your local library and then tell me why libraries are desperate for funding. I can't understand it — these are our number-one community institutions for information. Since we keep hearing that this is the information age, wouldn't you think they'd be a political top priority — the center of every neighborhood, where people go for the latest news?"

To the extent that libraries can continue to serve this purpose, it will be because citizens affirm the importance of their role in the face of the encroaching commercial orientation to information and knowledge. "Libraries in America are situated on the boundary between the market and the polity," says Peter Lyman, university librarian at the University of California, Berkeley. They provide "free access to knowledge in order to fulfill the public interest in education and democratic participation."

Licensing Mona Lisa

Like the nation's universities and libraries, museums also face a barrage from commercial forces. Seeing the potential for profits, a number of private firms are now clamoring to gain digital rights to museum collections. But no consensus has emerged in this uncharted field about how profits should be distributed among the museums that own the artwork originals, the artists who created them, or the new firms poised to collect and disseminate works in this new format.

"It's open season on museum collections," laments Janice Sarkow, a curator at Boston's Museum of Fine Arts, who says her museum — the oldest in the nation — has been inundated with requests from aggressive private firms representing on-line encyclopedias, advertisers, and CD-ROM vendors for ownership rights to digital reproductions of the museum's holdings. She explains, "Most of these vendors don't even know what types of rights they're asking for." The firms often seek the broadest ownership rights they can wrangle so they can retain control as the technology evolves.

By encoding a work digitally, the owner could send you a good, high-resolution copy of the Mona Lisa over the Internet or display it on a large flat screen in your living room. Microsoft tycoon Bill Gates is so confident of future trade in digital imagery that he has built large flat screens into many of the walls of his enormous home outside Seattle expressly for the purpose of showing digital art. When he wants to travel to Africa, he says, he can call up images of African art and artifacts at will in preparation for his trip. Gates is so convinced that owning digital rights to artwork can be a "gold mine" that he has been buying up the rights for nearly a decade. His little-known "other" company — the privately held Corbis Corporation — has for years been quietly making deals with art museums and other institutions to buy the digital rights to the art, photography, and other artifacts in their collections.

Through Corbis, Gates's digital collection has grown to include an astonishing 1.3 million images, with rights to at least 16 million more. He already boasts more works than fill the exhibition space of all but the world's largest museums — including rights to everything from the complete works of photographer Ansel Adams to one of the largest collections of Impressionist paintings in the world.

Corbis representatives like to say that the company is becoming a "digital Alexandria," the electronic equivalent of the great library that Ptolemy I built in the fourth century B.C. to hold the entire sum of human knowledge. The Seattle-based firm has already succeeded in positioning itself as a "comprehensive digital-image archive" that "licenses and purchases carefully selected sets of image collections from around the world" on topics ranging from art and architecture to world geography, people, and cultures.

From one perspective, Corbis is making an offer that is difficult for museums to refuse. The current market value of digital rights to artwork is still unclear, but Corbis is offering the museums a share of the revenues that will accrue from their pay-per-use dissemination. And of course, the deal in no way hampers the originals, which still reside in the museum collections.

But for all its potential power to disseminate these works widely, Corbis is not a museum. It is a private, for-profit business, owned by the world's richest man. Is it farfetched to worry that rights to the Mona Lisa, or to Robert Mapplethorpe's photographs, could just as easily be sold to a business tycoon with more overtly political interests — such as the censorship of certain works — with no public input? As on other fronts of the knowledge wars, we risk losing our rights to a valuable shared cultural legacy before we have fully thought out the consequences.

Museums could easily get into the same business themselves, fulfilling their public mission by disseminating digital imagery in cyberspace. Some museums, as part of the recently formed Art Museum Image Consortium (AMICO), are starting to do this. In just one of many examples, the National Gallery of Art in Washington, D.C., has encoded its entire collection on CD-ROM. A new Micro Gallery offers visitors a chance to browse through the museum's artwork on one of seventeen high-resolution color monitors, retrieving works by subject, artist, date, or place of origin. CD-ROMs holding digital images from a number of museum collections are already on the market for viewing on home computers.

The ability to browse on-line through a given artist's entire opus, or otherwise cull through the world's art collections is an enticing vision indeed. The central question at issue is: who should profit from the widespread dissemination of these digital images — the museums, the artists (or their estates), or a new set of firms that collect and transfer the works into this new format? If the situation at the nation's libraries is any guide, the latter group has a good chance of prevailing, barring some kind of collective intervention.

At Boston's Museum of Fine Arts, Sarkow believes that "digital imaging technologies pose some of the most important questions museums face today." Museums must be careful, she says, not "to relinquish open-ended rights" to their collections. As she puts it, "We feel very strongly that we are preserving a legacy."

Bowing to the concerns of museum curators and administrators, Corbis has increasingly been forced to enter into nonexclusive arrangements with museums, holding open the important possibility of competition — or even the prospect that the museums or other nonprofit players will offer an alternative to the commercial dissemination of digital images in the future.

Katherine Jones Garmil, a curator at Harvard University's Peabody Museum and the program director of a nationwide group called the Museum Computer

Network, describes the current environment as one of scrambling arrangements, transition, and flux. "Until recently, interest in technology within museums was found in small pockets — one particular curator or someone on a special project," she says. "But now these are really burning issues for museums and we're seeing much more widespread interest." With the advent of widely accessible computer imaging tools, "everyone seems to want to be part of it." Garmil notes that exciting possibilities are emerging that could allow much wider access to art through digital images. But the pitfalls are clear as well. As she puts it, "We just have to be careful not to simply sell our cultural heritage to the highest bidder."

Even a cursory review of the status of our storehouses of knowledge — the university, library, and museum — finds institutions whose public-spirited mandates are jeopardized by relentless commercial pressures. In each case, the expansion of the notion of ownership into the conceptual realm undermines these institutions' time-honored ability to ensure widespread and unobstructed access to the intangible knowledge assets they hold.

8

Panning for Drugs

JUNE 1996, QUITO, ECUADOR. To Valerio Grefa, U.S. Plant Patent 5,751 was both an outrage and a godsend. In the spring of 1996 the Ecuadorian parliament was on the verge of ratifying an intellectual property treaty with the United States. Grefa, then head of COICA (pronounced *koh-ee-ka*), an organization representing some four hundred indigenous groups throughout the nine nations of the Amazon Basin, believed that Ecuador's government was selling short the rights of indigenous peoples to their land and traditions. Among other provisions, the treaty would require Ecuador to recognize and honor U.S. patents, including patents on traditional native healing methods and Ecuador's wild plant varieties.

Luckily for Grefa, American activists affiliated with an international group called the Indigenous People's Biodiversity Network had been combing through patent databases in the United States and other developed countries. They were searching for controversial ownership claims, and when they discovered U.S. Plant Patent 5,751, they passed word to COICA. Grefa called the patent "a true affront to the culture of our peoples." But he also recognized an opportunity: the patent represented just the kind of explosive case COICA needed to galvanize opposition to the U.S.-Ecuadorian accord.

From a small office in downtown Quito, Grefa penned a press release that would attract worldwide attention to the case and to the rights of indigenous people. Skirting the fact that the patent was ten years old, Grefa publicized the case in incendiary tones: a U.S. citizen had obtained private ownership rights to a plant species that is the key ingredient of *ayahuasca*, a sacred hallucinogenic drink used for centuries in religious ceremonies throughout the Amazon. To Grefa the patent was nothing short of sacrilege. He saw no reason why anyone, especially a U.S. citizen, should be able to own rights to a living piece of Amazonian heritage. He likened it to someone from the Amazon coming to America and claiming exclusive rights to communion wafers.

Putting the issue in a neocolonialist frame, Grefa contended that the U.S. patent represented "an act of piracy" characteristic of many similar efforts to appropriate the plants and traditional knowledge of indigenous peoples. And, if the U.S.-Ecuadorian accord were ratified, he warned accurately, the "ayahuasca patent" would become valid in Ecuador. His skillful depiction conjured visions of greedy U.S. executives descending upon the Amazon to collect royalties from worshiping natives.

"The person who has had the audacity to patent this ancestral knowledge," Grefa wrote, "is Mr. Loren Miller, owner of the biopiracy company 'International Plant Medicine Company.'" COICA's council, Grefa announced, had decided to ban Miller from entering any indigenous community in the region, branding him "an enemy of indigenous peoples in the nine states of the Amazon Basin." The press release aimed a thinly veiled threat of violence at Miller and his associates, proclaiming that COICA "will not be responsible for their physical safety should they choose to ignore this resolution."

How had a lone U.S. entrepreneur engendered such antagonism? And how had he come to own a plant variety widely held to be sacred? As it turns out, Loren Miller, deftly portrayed by Grefa as a blaspheming enemy of indigenous peoples, more closely resembles a hapless tourist sightseeing on the battleground in a shooting war. But, as Grefa was well aware, neither Miller's personal circumstances nor his ostensibly benign intent alters the volatile nature of the issues raised by this case. A global gold rush is on: the prospectors are the world's pharmaceutical companies and, like Miller, they are panning for drugs.

Multinational pharmaceutical companies make up a staggeringly large economic bloc, with annual worldwide sales in excess of $250 billion. Not only is drug making a huge endeavor, it is also one of the world's most profitable manufacturing sectors, in itself strong testament to the power of knowledge assets in the global economy.

Yet drug pricing is a tricky business. Pharmaceutical firms naturally want to defray their enormous research and development costs. Nonetheless, once a drug is developed, its manufacturing costs are low, as they are in the software industry — in fact, such low costs are a classic feature of a knowledge-based industry. In the pharmaceutical field the debate over costs, pricing, and profits ricochets poignantly because serious medications offer a textbook example of what economists call inelastic demand. When people are sick, they will bear just about any cost to relieve their suffering.

Against the backdrop of the vast global enterprise to develop and market drugs lies the important fact that since the dawn of civilization, healers have distilled treatments and tonics from natural sources. Around 400 B.C., for example, Hippocrates documented the analgesic properties of willow bark. At the end of the nineteenth century the German pharmaceutical company Bayer refined this ancient remedy to yield the all-purpose pain reliever it dubbed aspirin. In 1928, Alexander Fleming extracted penicillin from a natural mold; more recently, the antibiotic streptomycin was developed from a microbe in a soil sample.

Even today, when laboratories can synthesize extremely complex compounds from their constituent chemicals, nature is still the best teacher. Researchers are more cognizant than ever that many insects, plants, and other natural substances have evolved with powerful chemical defenses to repel predators. And, thanks to the latest technology, the task of screening the natural world to identify prospective drugs has become easier and more accurate. As Peter Hylands, chief scientist at the pharmaceutical firm ESCAgenetics Corporation, has put it: "I think people are really waking up to the fact that 25 percent of all prescription medicines are still derived from plants, and yet only 10 percent of the world's plants have ever been looked at with modern screening methods." The result: an intensified quest for dominion over this global genetic treasure trove, with pharmaceutical firms systematically moving to catalog and claim rights to the medically interesting substances found in wild plants, fungi, microbes, and animals. Loren Miller's patent is merely one tiny piece — albeit a politically provocative one — of a very big picture.

Today Miller expresses only bitterness at having been drawn into a politicized diplomatic scuffle. First of all, he contends, COICA chose for political purposes to depict his patent as covering the sacred ayahuasca brew itself, even though he has always been interested in only one chemically active ingredient of it. "COICA completely misconstrued my work," Miller says. "It's like they're saying that I patented pizza when all I did is patent a rare type of cheese." With some justification, he complains that his tiny International Plant Medicine Company is a convenient, relatively powerless, scapegoat. In response to Grefa's belligerent charges about his work, Miller gripes, "COICA found themselves a horse that jumps and they don't care if the story is exaggerated or even just plain wrong."

Despite Miller's seemingly brazen act of securing a patent on a sacred plant, Grefa could hardly have found a less likely poster boy for corporate colonialism. As Miller tells it, a native healer in Ecuador gave him a cutting of the vaunted *Banisteriopsis caapi* vine — known in the Amazon as the "vine of souls" — back in 1980, when Miller was a graduate student in psychology at Stanford University. Miller was fully aware that the plant was the key ingredient in ayahuasca; that was what had drawn him to the region. In his graduate work, he had latched on to the idea that the psychoactive substance in the vine might be useful in psychotherapy or in the treatment of sleep disorders.

By his own account, Miller, a self-described hippie at the time, was fascinated by hallucinogens and the wisdom of Latin American shamans. In this regard he was following a well-worn path. Scores of seekers from the United States and Western Europe, looking to "expand their consciousness," had ventured to Central and Latin America and Asia to sample the native pharmacopoeia. Perhaps most prominently, the novelist William Burroughs made a pilgrimage to Ecuador in the 1950s to try ayahuasca, also known as *yage*, finding an enthusiastic cult readership in his book, *The Yage Letters.*

On his journey, Burroughs had taken along his friend and former Harvard classmate Richard Schultes, a near-legendary ethnobotanist who, a generation before, had traipsed around the world studying exotic and psychoactive plants and sending them back to Harvard for further study. Miller counts Schultes as a major influence on his career, and he rightly notes that taking samples from wild plants in exotic locales is part of a long-standing scientific tradition; the collection gathered by Schultes and others at the Harvard Herbaria runs to the tens of thousands of varieties. And many plant collections at the world's preeminent botanical gardens are much larger. The picture gets considerably more complicated, however, when the plants' active ingredients are privately owned and lead to enormous profits.

Genetic Resources: A Vital Input

Pharmaceutical firms learned long ago that they can reap handsome profits by claiming exclusive ownership of the formulas of naturally occurring ingredients. Substances derived from Madagascar's rosy periwinkle flower, for instance, yielded the drugs vincristine and vinblastine, used against Hodgkin's disease and juvenile leukemia, respectively. These drugs have, for many years, earned the patent holder, Eli Lilly & Company, some $160 million annually. The immunosuppressant drug cyclosporine, administered to transplant patients to suppress rejection of donor organs, is derived from a species of fungus. It earns the firm Novartis roughly $100 million each year. An exotic fungus yielded the cholesterol-lowering drug Mevacor; it has earned its owner, Merck & Company, some $700 million annually since 1990. And the quest continues. The U.S. Food and Drug Administration approved a substance discovered in the bark of the Pacific yew tree, known as taxol, as a treatment for ovarian cancer in 1992 and breast cancer in 1994. With more cancer-fighting uses being developed, taxol promises a vast market for patent holders, such as Bristol-Myers Squibb.

Even a cursory look at drug development reveals that the world's storehouse of plants and other genetic resources is a vital input to pharmaceutical companies, offering a kind of fuel to drive the industry. Given the importance of those natural — and formerly shared — global resources, it is not surprising that international tensions would surface about how to assess their value and apportion their bounty. The situation is especially tense because the bulk of the globe's biodiversity is consolidated in the Southern Hemisphere, probably because the last glaciation in the Northern Hemisphere, 10,000 years ago, reduced biodiversity greatly. Today, many of the globe's species-rich tracts exist in less developed countries, far from the developed countries where almost all of today's pharmaceutical firms are headquartered.

Against such a backdrop, Loren Miller's case takes on new meaning. In the past decade, an international movement has emerged seeking a more equitable apportionment of the value represented by natural and indigenous knowledge resources. Of course, when Miller received his patent in 1986, it was inconceivable that he would become a lightning rod in an international fight over intellectual property claims. At the time, as he recalls it, he was excited about his research and sought to ensure that he would profit from his findings. To avoid sharing with Stanford a patent or any royalties that might accrue, Miller dropped out of graduate school in the fall of 1984 and applied for a patent on his own. In his garage he established what he calls a "virtual corporation": the now-infamous International Plant Medicine Company.

At the time, Miller says, he needed a patent simply to attract funding for his research. From a commercial standpoint, his contention is probably accurate. From tiny start-ups to multibillion-dollar pharmaceutical firms, private research enterprises secure patents on the material they hope to develop into effective drugs in order to attract investment and stymie would-be competitors. "Without a robust patent system," says Patricia Granados, a patent lawyer at the Washington, D.C.–based firm Foley and Lardner, "it is questionable whether biomedical industries could obtain the investment money needed for research and development." As Granados explains, "Corporations aren't going to invest their money in anything unless they can get a proprietary position."

Patents are so central to the pharmaceutical industry that intellectual property specialists frequently cite it as a rationale for the entire patent system. Sitting behind a mound of blue file folders containing pending patent applications, Deputy Commissioner Charles Van Horn of the U.S. Patent Office offers the classic example of a pharmaceutical firm trying to innovate. "On average, drug companies now spend $250 million to bring a drug from conception to approval by the FDA. If you don't have some way to say to that company that their work will be protected, you put them at tremendous risk." Without patent protection, Van Horn contends, most companies would shun such risks, and "the public would never benefit from the fruits of that kind of research and many great ideas would never be realized." In other words, many drugs would never get to market.

Miller, a novice at the patent game, protected his "find" by obtaining a plant patent, which has been available since Congress passed the Plant Patent Act in 1930. Administered by the U.S. Department of Agriculture, this type of patent provides that for twenty years no one in the United States can grow or use Miller's plant species without his consent and, presumably, payment of royalties. Yet far more common among pharmaceutical firms, and even more powerful, is the protection afforded by a utility patent, administered by the U.S. Patent and Trademark Office. Applicants who can characterize the chemical makeup of a particular plant compound can be awarded exclusive rights over a whole new class of medicines with little more than the suggestion — not necessarily proven — that the substance will be safe and effective for a particular medical use.

Miller hopes to eventually obtain a utility patent. In any case, he is on solid ground when he contends that he was merely following commonly accepted practice in patenting the Ecuadorian plant variety. But it is a defense that largely makes Grefa's point. Each year, according to one estimate, drugs derived from plants in the Southern Hemisphere earn the pharmaceutical industry some $32 billion. Historically, these resources have been freely shared as part of the public domain. Ecuador and other countries where these plants are found have earned no share in the profits derived from them. The intellectual property protection garnered by pharmaceutical firms has assured not only that the drugs reach the market but that all revenues accrue directly to the technological titleholders.

Today a growing chorus questions the fairness of this arrangement. Companies in the developed world hold the overwhelming majority of patents on naturally occurring ingredients from the Southern Hemisphere. One study, published in 1997, found that multinational corporations held 79 percent of all utility patents on plants, and northern-based research institutions and universities controlled 14 percent. Parties in the United States owned more than three-quarters of this lion's share of the world's patents. Parties in developing nations hold so few patents they are virtually unrepresented in the statistics.

As if these disparities were not enough, Miller's case illustrates an additional politically explosive dimension: researchers are frequently alerted to plants with medicinal potential by local healing practices, indigenous traditions, and lore. Several small pharmaceutical firms in the United States have been founded expressly to tap this wellspring of shared local knowledge. California-based Pharmagenesis, for instance, brings the plant-based medicines of Asian native healers to the global market. Shaman Pharmaceuticals, also based in California, looks mostly to healers in Central and Latin America for leads on new drugs to patent and market.

Of course, the usurpation of traditional knowledge, like the appropriation of the plants themselves, has long-standing colonialist roots. A good example is the drug curare, which for centuries the Makushi Indians of Guyana put on their arrowheads to incapacitate their quarry. The Makushi people divulged the secret of curare's source to British explorers in the early 1800s, and the result, for the developed world, was an anesthetic and muscle relaxant that still generates millions of dollars in sales. The natives never received a shilling of compensation or royalties for sharing the knowledge. Neither did the Amazonian peoples who used the bark of the South American tree *Cinchona officinalis* to treat fever for generations before Europeans, calling it quinine, adopted it for fighting malaria.

In a more recent example, Minneapolis-based MGI Pharma developed a drug called Salagen to alleviate the symptoms of xerostomia, or dry-mouth syndrome, in which a person is unable to produce saliva. The drug's active ingredient, pilocarpine, is extracted from a plant native to northeastern Brazil. To capitalize on the indigenous people's knowledge, MGI Pharma need only have referred to the plant's native name, *jaborandi,* or "slobber-mouth plant." And yet MGI Pharma paid nothing to the native Brazilians who have known about the plant's therapeutic properties for generations.

Such manifest inequity led a United Nations–sponsored report in 1994 to brand transnational pharmaceutical firms "biopirates," contending that private companies should not be allowed to stake exclusive claims to indigenous knowledge that has been passed down and freely shared over many generations. As the report noted, roughly 80 percent of the world's people depend on traditional knowledge and wild plants for their medicinal needs. Local healers administering traditional cures often have considerable knowledge about the effects of various plants, even if they don't know their molecular makeup and would never think to patent their medicinal properties. Profitable drugs that annually earn pharmaceutical firms some $5.4 billion can be directly linked to the usurpation of traditional local knowledge, the report claims.

Today sentiment has begun to swell in opposition to the uncompensated usurpation of indigenous knowledge. Jason Clay, former research director of the nonprofit group Cultural Survival, says simply: "It's a question of intellectual property rights. People whose medical lore leads to a useful product should have a stake in the profits." As Clay puts it, "Unless we return some of the profits to them, it's a kind of theft."

As Clay's view begins to find a degree of mainstream acceptance, it offers further insight into why Miller's patent struck such a nerve in Ecuador. Not only did Miller claim to personally own a holy plant, he claimed private control over a plant variety whose properties were already well known and appreciated by indigenous communities.

In highlighting Miller's case, Grefa used a kind of tabloid tactics to intensify anti-American sentiment in Ecuador. But, hyperbole or not, Grefa's campaign to highlight the ayahuasca issue produced results. It built on momentum gained in the spring of 1996, when a coalition of indigenous representatives won an unprecedented number of seats in the Ecuadorian parliament and in local governments. By August, concern over the Miller case had escalated. A gathering of indigenous Amazon leaders condemned the ayahuasca patent as an act of biopiracy. Activists, citing the patent and other gripes against U.S. and multinational firms, even briefly occupied Ecuador's legislature, successfully blocking the parliament's effort to ratify the treaty. In this highly polarized climate, despite pressure from the U.S. embassy in Quito, the legislature sent the treaty back to the negotiating table.

Ecuadorian and U.S. officials, as well as activists for indigenous people around the world, watched, astounded, as Grefa's tactics played a key role in scuttling a trade accord between Ecuador and the United States that had been in the works for more than three years. Antonio Cobo, an undersecretary in Ecuador's Ministry of Foreign Trade, for instance, voiced his government's support for ratification, arguing that the treaty would bolster trade between the two countries. But that view could not carry the day. As Cobo complained, "The bilateral agreement has unfortunately become politicized."

Grefa scoffed at the complaint. "This is a political issue and should be understood as such," he retorted. "This is an issue of sovereignty, recognition, and self-determination for our people and our resources." According to Grefa, COICA and other indigenous organizations want "the ancestral knowledge of indigenous people to be recognized as a contribution to human progress."

It didn't matter that the International Plant Medicine Company was tiny, that Miller's patent was nearly a decade old, or that he professed no intention to enforce it in Ecuador, even if the U.S.-Ecuadorian accord were ratified. Miller's case put COICA's concerns in clear neocolonialist relief: U.S. companies were actively appropriating ingredients of indigenous peoples' shared heritage — even sacred ones — and claiming to own them.

To pharmaceutical industry representatives and many other international observers, it was ironic that it was Miller's tiny and somewhat innocuous case that caused the fuss. After all, virtually every major pharmaceutical company in the world has ownership claims over natural genetic resources that make Miller's plant patent seem minuscule by comparison.

Bioprospecting Comes of Age

Who, if anyone, should be able to claim ownership rights to the globe's genetic and cultural inheritance? Tensions have been brewing in earnest over this question for more than a decade, but a watershed came in the autumn of 1990, when Cornell entomologist Thomas Eisner helped forge a deal between the giant pharmaceutical company Merck and a fledgling nonprofit organization in Costa Rica called the Instituto Naçional de Biodiversidad, or INBio. At the time Eisner was studying insects in Costa Rica. That country's rain forest, a 5,000-square-mile tract of jungle covering roughly a quarter of the nation, holds what may be the planet's densest and most diverse array of plant and animal species. It is believed to support between 5 and 7 percent of the globe's total number of remaining species, including an estimated half million species of insects, plants, fungi, mammals, reptiles, and birds, and an unknown number of diverse bacteria and viruses.

Eisner was impressed with INBio, created in 1989 to study and preserve Costa Rica's diverse biological heritage. In 1990, INBio embarked on an ambitious program to catalog the country's species in a computer database and note their distribution, abundance, and potential uses in medicine, agriculture, and industry. To Eisner, the massive project was a boon to science and in particular to his own research, concerning how insects manufacture and use chemical substances to attract prey and repel predators, research that had drawn much interest from pharmaceutical companies. Eisner naturally thought that INBio's cataloging project might also attract corporate funds to expand and underwrite its inventory efforts.

Eisner stresses that his interest was fueled largely by the rapid disappearance of species — and their genetic material — from the face of the earth. By the 1980s he had already experienced the widespread destruction of insect habitat, and he knew how grim the overall picture was. Tropical forests, predominantly in Central and South America and Southeast Asia, contain from 50 to 90 percent of all species, including two-thirds of all higher plant species and up to 96 percent of insect species. But at current deforestation rates, up to 8 percent of all rain-forest species will become extinct by the year 2015, and 17 to 35 percent of all species will be threatened with extinction by 2040. For naturalists like Eisner, the stark and accelerating extinction rate presents a race against time. After all, scientists predict, the 1.5 million species taxonomists have named so far represent only 5 to 30 percent of the globe's varied life forms.

Eisner recognized that deforestation means not only the destruction of natural habitats and the extinction of species but also the loss of potentially profitable drug ingredients. Toward this end he wrote a landmark paper in

1989 promoting what he dubbed chemical prospecting, now more commonly referred to as bioprospecting. More pieces fell into place when Eisner connected with Paul Anderson, a colleague's former graduate student, who had become the senior vice president for medical chemistry at Merck. After negotiating for almost a year, Eisner and several other enthusiastic colleagues persuaded the pharmaceutical giant to cut a deal for access to Costa Rica's biodiversity. The agreement heralded a new era of bioprospecting.

In the 1991 deal, Merck agreed to give INBio a two-year research budget of $1.2 million and an undisclosed share of royalties (eventually reported to range between 1 and 3 percent) on any resulting commercial products. INBio, a private, nongovernmental research institute, would provide exclusive access to some ten thousand Costa Rican species, gathering samples of the country's wild plants, insects, and microorganisms for Merck's drug-screening program.

From almost any standpoint, Merck had struck a good deal. Given Charles Van Horn's figure that pharmaceutical companies invest an average of $250 million to bring a new drug to market, the entire INBio deal barely amounted to "loose change," as one observer put it. The comment holds especially true considering that Merck sold nearly $20 billion of drugs and medical products in 1997. Even in 1991, when it finalized its agreement with INBio, the company's $8.6 billion in sales considerably exceeded Costa Rica's entire $5.2 billion gross national product. Then as now, Merck's three top-selling drugs each had annual sales rivaling the Costa Rican government's annual $1 billion budget.

At the rate Merck paid for access to Costa Rica's share of the world's biodiversity, the entire planet's storehouse of genetic resources would fetch just $20 million. Sample by sample, Merck paid roughly $100 per species, certainly a modest sum considering that a winning sample — much like a winning lottery ticket — can result in a $1-billion-a-year drug.

In addition, as Merck representatives surely must have realized, they could justify the INBio deal solely on the basis of the upbeat publicity it generated. In an industry that spends some $10 billion a year promoting its products in the United States alone, Merck looked like a corporate angel helping to save the rain forest — and all for less than the company might have spent to promote a single new drug.

The Merck-INBio agreement met with nearly universal acclaim when it was announced. Environmental groups, Costa Rican leaders, biotechnology and pharmaceutical firms, and U.S. government officials all hailed the venture as a way to help preserve the precious storehouse of biodiversity in Costa Rica's rain forests. Eisner called it "a win-win situation." Costa Rica would benefit from conserving its natural resources, and at the same time the agreement would "protect the proprietary rights" of the drug industry. Ownership rights aside, Merck was not shy about emphasizing the environmental angle. Landowners in Costa Rica are often interested in cutting down the forest to have more farming, as Georg Anders-Schönberg, then Merck's director for natural products chemistry, explained to a reporter from *Environment* magazine. If they are involved, he said, they become more protective of the forest.

Jessica T. Mathews, then vice president of the World Resources Institute (WRI) in Washington, expressed the prevailing view among U.S. environmentalists, noting that such agreements represented "the most efficient way to achieve environmental ends" by making resource conservation more profitable than resource exploitation. According to WRI's calculations, if INBio earned 2 percent of the income Merck derived from the sale of twenty average products — a plausible guess, given some ten thousand samples — INBio could amass more money than Costa Rica earns from coffee and bananas, its two leading exports.

Lost in the hoopla was any recognition of a momentous change that had taken place. Wittingly or not, the Merck-INBio contract established biological resources as a recognized component of international commerce. Genetic resources that had virtually always been a shared public good were now sanctioned as a commodity.

Merck and INBio have repeatedly renewed their arrangement, and Merck scientists say they have already isolated several compounds from INBio samples that show promising biological qualities. However, John Doorley, a representative for Merck, says that the deal has yet to result in any actual drugs under development. Doorley stresses that drugs routinely take twelve years or more to reach the market and that it is early yet to expect much.

INBio has built on its success with Merck, concluding similar agreements with other companies, including Bristol-Myers Squibb, which bought rights to a different set of samples from the Costa Rican rain forest. INBio hopes to complete its species inventory within the next decade, subsidizing the massive effort through such agreements and, eventually, royalties from drugs and other products derived from these sources.

Meanwhile, the international trade in genetic resources has grown dramatically. Pharmaceutical companies have scoured the world's jungles, herbariums, and folk pharmacopoeias in search of exclusive access to the biological information they contain. In a major public survey of the trend, conducted in 1995, the Rural Advancement Fund International counted some fifty-one companies involved in bioprospecting agreements. Some of these firms had entered into as many as five separate contracts, covering access to everything from the indigenous plants of Suriname to the marine organisms of Micronesia. The list of players includes almost every major drug company, including Bristol-Myers Squibb, Eli Lilly, ESCAgenetics, Glaxo Wellcome, Johnson & Johnson, Merck, Pfizer, Pharmagenesis, Phytera, PhytoPharmaceuticals, Rhône-Poulenc Rorer, Shaman Pharmaceuticals, SmithKline Beecham, and Upjohn.

Most of these agreements represent variants on Merck and INBio's one-to-one arrangement between a drug or biotechnology company and a group or nation, usually in the Third World. But the agreements don't exclusively involve developing countries. In the United States, for instance, the San Diego–based pharmaceutical firm Diversa arranged a deal with the U.S. National Park Service to collect and patent microorganisms from Yellowstone National Park's geothermal areas. The United States explicitly prohibits the patenting and commercial exploitation of national park resources, but the park service and Diversa circumvented this by agreeing to consider these resources as a form of information, a category of resource never considered in the federal statutes.

John Varley, Yellowstone's director of resources, says that the agreement represents the shape of things to come. He sees widespread research interest in the genetic resources in national parks and says that Yellowstone stands to receive royalties amounting to 10 percent of sales of any drugs or other products that Diversa might develop. The Diversa agreement may be a good way of garnering additional revenues, but, given the protected status of the park, many believe the matter should be fully debated in a democratic process. As a result, the Diversa agreement has already drawn a legal challenge from two U.S. nonprofit groups, the Edmonds Institute and the International Center for Technology Assessment. "We didn't preserve Yellowstone for corporate purposes," says Beth Burrows, director of the Edmonds Institute, who calls the deal "a theft of our national heritage."

If the Merck-INBio prototype helped spawn genetic resource agreements around the world, it also played a key role in helping to codify the commercial trade in genetic resources into international law. The issue of ownership of these resources was a prominent element of the Convention on Biological Diversity, which entered into force in December 1993. Ironically, during the 1992 world environmental summit in Rio that hammered out the accord, U.S. drug companies and organizations, including the Industrial Biotechnology Association and the Association of Biotechnology Companies, worried that developing countries would win too much control if a treaty formally sanctioned worldwide global commercial trade in genetic resources. Consequently, they successfully pressured President Bush to withhold U.S. endorsement for the entire treaty, despite its worthy goal of protecting biodiversity.

Although the Clinton administration subsequently signed the accord, the concerns of the pharmaceutical and biotechnology industries were badly misplaced. The Biological Diversity Convention, now ratified by 169 countries, has served both industries exceedingly well. The treaty does recognize the sovereign rights of states over their natural genetic resources, but the treaty allows multinational corporations to freely patent genetic materials. Despite a hortatory passage encouraging "the equitable sharing of benefits arising from the utilization of such knowledge," it offers no specific guidelines for compensating nations or indigenous peoples responsible for nurturing, using, and developing biodiversity worldwide.

Global acceptance of a regime favoring private claims over natural genetic resources was further solidified in the GATT negotiations in the early 1990s. The piece of these negotiations known as the Trade Related Aspects of Intellectual Property (TRIPs) agreement obligates all signatories, including developing countries, to adopt minimum intellectual property standards allowing for the use and ownership of plants and microorganisms. Historically, the issue of intellectual property has been largely left up to individual governments. Most developing nations chose not to recognize patents on food, pharmaceuticals, or products that met other basic human needs, for instance. Now, however, under the guise of "harmonizing" intellectual property laws between nations, the agreement drives the developing world closer to the commercial view of genetic resources adopted by the United States and other developed nations. With the TRIPs accord, developing nations face trade sanctions if they fail to adopt strong intellectual property protection for the private control of these resources.

The net result of these international treaties is to greatly expand the power of corporate titleholders. Taking advantage of the huge changes still under way, a handful of corporations has quickly staked far-reaching claims to the globe's living organisms, amassing proprietary knowledge assets that give them a powerful advantage in the quest to develop drugs, treatments, and other biotechnology products.

Valuing Indigenous Knowledge

As drug companies lay claim to the world's remaining store of biodiversity at an intensifying pace, most observers agree that the world's genetic resources hold enormous commercial potential. Some go so far as to contend that, taken together, they will prove to be the most lucrative asset on earth. That idea raises a host of thorny problems. Perhaps the most immediately vexing issue is: who has the legitimate right to sell natural genetic resources?

On the one hand, many industry representatives resent having to compensate anyone for access to living resources in the first place. Making the classic case for the pharmaceutical industry, Harvey Bale, former vice president of the Pharmaceutical Research and Manufacturers of America, emphasizes that making cures from wild plants is a process of invention; consequently, Bale says, the inventors are entitled to broad rights over their raw materials. As for royalty-sharing arrangements, he says, "It's as if, just because the American Indian is there and gives inspiration to John Wayne movies, then 20th Century Fox owes them royalties." He continues, "It ain't the Indians of the Wild West who are the creators of those films."

Taking a diametrically opposite view the philosopher Mark Sagoff argues that private ownership of the world's genetic resources is inappropriate precisely because they are *not* inventions. Sagoff believes that while these resources can clearly enable invention, they are not inventions themselves. An understanding of that fact, Sagoff says, will ultimately persuade the world to retain its living natural resources as a pooled asset. Any other outcome, he argues, is untenable. As Sagoff puts it: "If you lose the distinction between what is an object of nature and what is human design and invention, any sort of absurdity follows."

Still others argue that the Merck deal rightfully institutionalized the notion that genetic resources are a commodity under sovereign jurisdiction. As INBio founder Rodrigo Gamez states, "What we did is show that a developing country owns its genetic resources just as it owns its oil or minerals." The irony of Gamez's statement is that, despite INBio's close ties with the Costa Rican government, it has always been a *private* institute with a relatively tenuous claim over the raw materials it offered to Merck. As Christopher Joyce recounts in his 1994 book *Earthly Goods: Medicine Hunting in the Rainforest,* INBio finessed the point with what amounts to a kind of kickback to the government, agreeing to contribute 10 percent of Merck's up-front payment and 50 percent of any future royalties to the Costa Rican government's National Park Fund.

At the time of the original deal, several critics worried about the tenuous basis of INBio's claim to this biological material, but their voices were drowned out in the enthusiastic outpouring of support. Among the critics, Mario Carazo, a well-known lawyer in San José and codirector of the Costa Rican environmental organization Fundación Ambi, complained bitterly about INBio's private status: "They are a closed club, and they cannot sell the national patrimony." As for Merck's contribution, Carazo compared the drug company's payment to INBio to "the little mirrors the Spanish gave to the Indians." Noting that no royalties have yet accrued, Carazo warns that the history of Latin America is littered with broken promises.

Bolstering Carazo's neocolonialist assessment is that INBio gets no patent rights to the species it discovers on Merck's dollar. Like most bioprospecting agreements, the Merck-INBio deal stipulates that Merck will have all patent rights deriving from any active ingredients it receives in the arrangement. Contrast this to U.S. law, which affords even small universities some ownership rights over any patents granted to their faculty. Had Loren Miller kept his affiliation with Stanford University, for example, the school could have required him to share his patent rights. Yet, for housing, preserving, and stewarding its natural genetic resources, neither Costa Rica nor INBio has received an analogous arrangement.

In an equally powerful line of criticism, a number of analysts complain that the true conservators of healing techniques and wild species are often indigenous people, whom the Merck-INBio deal entirely ignores. Elaine Elisabetsky, a Brazilian pharmacologist and anthropologist, argues that native peoples have at least as rightful a claim on these resources as do private groups like INBio or even the Costa Rican government. Elisabetsky, among others, has charged that Merck purposely tried to avoid the "ethnobotanical angle" to keep indigenous people out of the equation.

Given this sharp disparity of views, it is not surprising that the growing trade in biomaterials has fueled North-South tensions. In the plethora of recent corporate deals, these conflicting and unresolved views portend serious trouble. The fact is, many parties can assert equally legitimate, overlapping claims to this natural bounty.

Take, for instance, the case of Walter Lewis, a plant biologist at Washington University in St. Louis. In 1993 the U.S. government launched an initiative to foster bioprospecting arrangements modeled loosely on the Merck-INBio deal. It established the International Cooperative Biodiversity Groups (ICBG) program as a collaborative effort by a number of federal agencies, including the National Science Foundation, the National Institutes of Health, and the Agency for International Development (USAID). Lewis, one of the first researchers to receive ICBG funding, proposed to search for medicinal plants in the highlands of Peru, particularly those used by the Aguaruna and Huambisa peoples. His problems began almost immediately.

Lewis negotiated a deal with the St. Louis–based pharmaceutical firm G. D. Searle to commercialize any medically significant biological materials his team might find. But working with the indigenous people proved daunting. Unfortunately for all involved, the attempt to negotiate an agreement turned into a bitter clash of cultures. Hoping to quickly settle the legalities, Lewis's team urged the tribes to sign agreements that offered them royalties if any of their treatments were successfully commercialized. This approach foundered when the two rival groups each claimed rights to the local plants and knowledge of their medicinal properties.

Lewis sought to overcome the problem by establishing a joint group that could speak for both tribes and dispense payments or royalties between them. But the effort was plagued by disagreement over who could legitimately represent the groups and what qualified as informed consent of the legal particulars. All of these points were of particular concern to Searle, which wanted clear, unfettered title to any drugs it might develop from the information. In urging the two indigenous groups to sign the legal agreement, Lewis and his team ultimately alienated both tribes, who complained that their views were being ignored and that they were being played off against each other.

Ultimately, the Aguaruna and Huambisa did collaborate — on a bitter appeal to the National Institutes of Health. "We energetically reject the lack of transparency, impositions, and manipulations of [Washington University's] team and demand that they immediately withdraw from Aguaruna and Huambisa territory," they wrote, calling upon the U.S. government to "correct and redress the aggressions made against indigenous peoples' rights in relation to Amazonian plants and expropriated indigenous knowledge." Sensitive to Third World perceptions in the aftermath of the fiasco, USAID in 1996 withdrew from the U.S. program altogether, which continues under the auspices of the other agencies.

The Lewis case underscores how difficult it can be to establish clear-cut ownership rights over indigenous knowledge. His team worked hard to reach a legal settlement. Often, though, would-be titleholders claim rights simply by delegitimizing indigenous knowledge systems altogether, as a recent case in India illustrates.

In 1993 two researchers from the University of Mississippi Medical Center applied for a U.S. patent on the use of turmeric to make wounds heal faster. U.S. Patent No. 5,401,504, issued in March 1995, claims a "method of promoting healing of a wound by administering turmeric to a patient afflicted with the wound." In asserting that their patent application covered a novel treatment, Suman Das and Hari Har Coyly argued that, prior to their work, there was "no scientific research" on turmeric as a healing agent for external wounds.

Even in their patent application, though, Das and Coyly acknowledged that turmeric is a commonly used traditional remedy. Known in many parts of India as *haldi,* the yellow powder from the dried root of the turmeric plant is widely applied to scrapes and cuts of children as a paste. But, as Coyly noted, "There are so many home remedies all over India, but are these scientifically valid or just gibberish? That's the point. We have used turmeric on patients. It has been clinically tested."

Das and Coyly found the U.S. Patent Office receptive to their argument. The agency will deny a patent to any proposed invention that has already been mentioned in published material anywhere but will consider the prior "public use or sale" of an invention as cause to disqualify a patent application only if that use has occurred in the United States. The prior-art rules of the U.S. patent code state specifically that a person is entitled to a patent unless it can be established that the proposed invention has already been "known or used by others *in this country* [emphasis added]."

The U.S. patent code derides as "folklore" those ideas and inventions passed on orally from generation to generation in other countries and explicitly ignores them, but there is little justifiable reason for this chauvinistic legal loophole. As Greg Aharonian, editor of *Patnews*, argues, "In this day and age, with information distribution tools like the Internet, with growing interest in nontraditional medicines and lifestyles, and with treaties like GATT encouraging a more global outlook," the loophole is "archaic." As Aharonian notes, in fact, a 1966 report to Congress recommended abolishing the Patent Office's geographic distinction. The recommendation has never been heeded.

Given the widespread knowledge of the turmeric treatment, a U.S. patent claiming exclusive right to it caused outrage in India. The Indian government formally challenged the patent, arguing that the use of turmeric as a wound-healing substance could in no way be construed as a novel "invention" because it was a well-documented traditional practice. Luckily for their case, Indian officials could produce published evidence — namely, a 1953 article in the *Journal of the Indian Medical Association* about the use of turmeric to heal wounds — that finally convinced the U.S. Patent Office to overturn the patent. But the turmeric "inventors" at the University of Mississippi are now seeking a narrower claim over medical uses of turmeric not specifically discussed in the 1953 article.

In the case of the turmeric patent and many like it, the emerging commercial trade in biomaterials and related indigenous knowledge represents a seemingly irreconcilable clash between two systems. Brigham Young University ethnobotanist Paul Cox notes that in traditional medicine, "knowledge is accumulated not by one researcher but by an entire culture through time." An understanding of the safety and efficacy of a drug often results from centuries of clinical practice. How could any would-be private owner equitably place a monetary value on that, not to mention deciding how to fairly distribute the proceeds?

The fact is, the loose, open system in which indigenous knowledge evolves and thrives is fundamentally threatened when confronted by a system of private ownership and monopoly patents. The intellectual property lawyer James Boyle has described the latter, now-dominant, global system as an "author-focused regime" that seeks to consider every incremental contribution as the invention — and hence the private property — of an individual author or creator.

Drawing the Line

The danger of heightened North-South tensions over natural genetic resources is cause enough for concern, yet equally troubling is the prospect that the private control of genetic resources will ultimately choke off future innovation. As Boyle warns, "The blindness of an author-centered regime to the importance of the public domain can lead to overly expansive intellectual property rights that deny future creators . . . the raw material they need to make new products." Ironically, partly in the name of protecting biodiversity in places like Costa Rica, we risk eroding another precious preserve: that of shared knowledge. Like the rain forest, this resource thrives on diversity and cross-pollination.

David Lambert, executive director of the American Seed Trade Association, contends that "patenting plants is no different than patenting any other kind of product." He couldn't be more wrong. Part of the problem is the commonly touted prospecting analogy. Compare gold and the seeds of a medicinal plant. Each unit of gold pulled from the earth has a market value based upon its size and purity, a value that represents only that of one discrete portion of the world's storehouse of gold. When you own a gold nugget, you make no intrinsic claim on any other units. The same holds true for crude oil or other traditional, tangible assets.

With a plant, however, every seed, every single cell, embodies a complete set of the knowledge that allows the species to propagate itself. Each cell thus has incalculable value, holding within it a universe of valuable information and an open-ended potential for financial exploitation. An exclusive ownership claim on the genetic information in a seed — even one pertaining to a particular medicinal use — needlessly threatens to prevent this knowledge asset from being investigated freely by others and thereby propagating new understanding.

For evolving and self-propagating systems like genetic resources, stewardship, not ownership, ought to be the primary concern. In this sense, the Costa Rican rain forest offers an apt lesson: even if clear-cutting trees in the rain forest is economically expedient in the short run by expanding available land for farming and allowing many houses to be built in a hurry, the forest's inhabitants will soon be left with a decimated landscape unless they harvest its bounty responsibly. Similarly, signs of trouble can be seen in the forest of knowledge. This is especially clear in university research programs, where many researchers have complained that secrecy and legalistic barriers prevent the exchange of information and biomaterials.

Interestingly, Merck has come to share these concerns, at least in the area of human genetic resources. Alan Williamson, Merck's vice president for research strategy worldwide, told the journal *Science* in 1997 that the company was becoming wary of a kind of layer-cake effect, as a tangle of multiple ownership claims threatened to confound the company's ability to create new products. As Williamson explained, while Merck researchers were accustomed to paying royalties on the many small patents going into each pharmaceutical product, the company "suddenly noticed that royalty claims were stacking up" on products to an unacceptable degree. Consequently, over the past several years, Merck has invested tens of millions of dollars in efforts to publish and share basic data about the human genome.

Despite this initiative, genetic information drawn from humans, like other genetic resources, is becoming fodder for bioprospectors. In perhaps the most publicized example to date, dubbed by the Associated Press as "a test case for the bio-age," the United States Patent and Trademark Office took the notion of patenting genetic resources into new terrain in the spring of 1995, issuing U.S. Patent No. 5,397,696 for an unmodified human cell line drawn from an indigenous person.

The fascinating episode grew out of research involving a dwindling, 260-member tribe of indigenous people known as the Hagahai. Living in a remote region of Papua New Guinea, the group had been isolated from the rest of the world until 1983, when a few tribespeople ventured out to seek help in combating malaria. An American researcher named Carol Jenkins began working with the Hagahai and, with help from colleagues at the National Institutes of Health, soon discovered that the tribe's blood carried a previously unknown, benign strain of human t-lymphotropic virus (HTLV-1). Normally the virus produces a severe form of leukemia, but the variant carried by the Hagahai left them unafflicted by the disease. Realizing that the strain of virus could be valuable in diagnosing adult leukemia and chronic degenerative neurologic disease, or even possibly in developing a vaccine, the U.S. researchers followed a now-standard procedure. They filed for a patent on the biological material, in this case a human cell line.

The rationale behind Patent 5,397,696 was similar to that behind the patent of any other genetic resource. The genetic information — in this case from an isolated human population — offered a possible aid in developing a medical treatment. The irony is that although the discovery may not lead to any commercial benefit, the NIH researchers say they sought the patent to preempt private firms from staking such a claim. As Jenkins has explained, if the NIH had not filed for a patent on the genetic information, "all profits from any commercial development conducted subsequently would go to the commercial company which exploited the discovery." Since the original specimen came from a Hagahai man in Papua New Guinea, the NIH researchers mentioned the tribe in the patent application. For her part, Jenkins said that if profits did accrue, she planned to return her share to the Hagahai.

Despite her promise, many in the international community found it deeply disturbing that the U.S. government was claiming exclusive ownership of genetic information derived from the cells of a person who was barely apprised of the situation. The Papua New Guinea government found the matter troubling enough to threaten to contest the patent in the World Court. Alejandro Argumedo of the Indigenous Peoples' Biodiversity Network, based in Canada, called it "arguably, the most offensive patent ever issued." Finally, in the fall of 1996, following an international storm of controversy, NIH formally filed to "disclaim" the U.S. patent.

Nonetheless, the incident threw into question the ethics of a U.S. government project known as the U.S. Human Genetic Diversity Project, which is affiliated with an immense international effort to map the human genome. This project, begun in 1991, is designed to collect genetic information from some seven thousand groups of people — mostly indigenous communities — around the world. Following the Hagahai episode, however, a 1997 report by a blue-ribbon panel at the U.S. National Research Council sharply criticized the program's plans. If human cell lines are to be patented, says this report, the effort should be conducted internationally under the auspices of a body like the United Nations. Undoubtedly the research could prove important in the discovery of new medical treatments, the report contends, but any human bioprospecting effort must be held to the highest standards of informed consent, profit sharing, and participation. Any information derived through such a program must remain in the public domain or at least in a trust under publicly accountable auspices.

These high-minded guidelines have slowed U.S. government efforts to gather genetic information from indigenous people, but they have had no obvious effect on private firms. Sequana Therapeutics at La Jolla, California, for instance, is reportedly spending $70 million to underwrite research on an isolated population on the South Atlantic island of Tristan da Cunha in hope of isolating the gene or genes responsible for asthma. The island's tiny population, descendants of people who settled it in the early 1800s, today suffer a high incidence of asthma. This is only one of many efforts of the firm that *Science* has branded "one of the more aggressive DNA prospectors."

Among the downsides of this effort is that the academic teams that have contracted with Sequana can't publish their valuable research on the island's inhabitants until the firm secures patent rights to any potential asthma genes or claims on related therapeutic products. As reporter Eliot Marshall puts it in *Science,* "Clearly this is one area of genetic studies where public resource sharing won't occur anytime soon. And it's an area that's likely to grow as many other companies jostle for access to DNA from genetically isolated groups that may point the way to disease genes."

There is little doubt that sophisticated genetic engineering techniques will lead to powerful new medical developments as the pharmaceutical industry taps the world's storehouse of genetic resources. But in today's commercial environment, widespread access to and robust exchange of genetic resources is by no means guaranteed. Genetic information, like most information, can be construed as a useful commodity to be bought, sold, and traded, but it cannot, in good conscience, be seen exclusively as such. Genetic information is important as a raw material through which to understand the world we share and the living organisms that inhabit it. Too much is at stake to limit future access to this natural heritage.

Part III

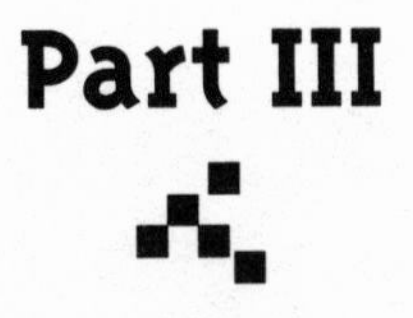

Common Knowledge

9
The Landscape of Invention

THE PRECEDING CHAPTERS offer dispatches from fights now under way in a variety of fields concerning access to and control of knowledge. They all follow a familiar pattern: as the value of information, concepts, blueprints, and formulas increases, players in the new economy of ideas actively claim rights to them. Types of know-how that had formerly been in the public domain become sought-after prizes in a procession of battles for private control. These knowledge assets often consolidate vast power into private hands with hardly a murmur from a largely unsuspecting public that, for the most part, remains barely aware of the situation.

Cumulatively, in the face of these developments, we are coming to recognize our collective stake in the matter. We all lose when physicians erode the ethic of sharing medical knowledge by claiming to privately own new procedures or insights about the workings of the human body. We all lose when copyright restrictions proposed by the information industry restrict the quality or volume of information public libraries can disseminate freely. We lose when the single genetic manipulation of a crop variety allows an aggressive biotechnology firm to claim rights to all related advances for the next two decades. Or when computer programmers levy so many demands for royalties on the basic building blocks of software code that they threaten to choke the field's development. We lose when pharmaceutical firms monopolize the biological materials that the world's researchers need to develop new disease-fighting treatments.

These kinds of broad private ownership claims are proliferating now because, in all these high-tech fields — including medicine, agriculture, biotechnology, software development, and the information industry — knowledge assets are rising in value compared to more traditional kinds of resources. Those heralding the Third Wave knowledge-based economy have characterized at least this part of the equation accurately. The world is witnessing a major shift to a new historical era in which knowledge assets play a driving role in economic growth.

Of course, technological improvements have always spurred economic growth, but never have they stood so close to the heart of economic life. As the widely accepted theory goes, the so-called First Wave preindustrial economy was marked by the private control of land. Agriculture, timber, hunting, mining — bounties that arose from the land — formed the foundation of economic well-being. During this preindustrial period, the big economic winners were landowners, who could use their dominion over real estate to control economic life within their fiefdoms. Wars large and small were waged to maintain or acquire power over territory.

Eventually, though, with the ascent of industrialization in the so-called Second Wave, the locus of power shifted, as Karl Marx astutely recognized, from control of the products derived from the land to control of the means of production. Of course, control of land and its bounty continued to be important, as it is today. But in the industrial economy the biggest winners were those capitalists — such as factory and railway owners — who controlled not just physical property but the tools to make and transport the mass-produced goods sought by a growing urban population.

Today the product-oriented manufacturing industries are being eclipsed as the control of knowledge and know-how moves to the vibrant center of the economy. The knowledge of how to efficiently use scarce material resources now assumes at least as much value as owning or controlling the materials themselves. The key in the emerging economy is ownership or control of the concept of production: the blueprint, formula, or essential information that may enable a sought-after development. Similarly, for some products such as software, marginal production costs approach zero: the value of these knowledge wares is almost entirely divorced from the costs traditionally associated with the production of tangible goods.

In the new economy the savvy players are fast becoming knowledge moguls — technological titleholders who control rights to new processes and products. This shift was captured clearly in *Fortune* magazine in 1994. As editor Walter Kiechel put it:

> To be sure, there are still industries in which having the factory confers a competitive advantage. But this is changing, fast, as more and more compa-

> nies realize that their edge derives less from their machines, bricks, and mortar than from what we used to think of as the intangible, like the brain power resident in the corporation. What kind of customer-delighting product can an outfit's people dream up, and who cares who actually puts the physical embodiment of their dream together?

Who cares, indeed! The dramatic shift Kiechel sketches has implications that are hard to overstate. An economy that prizes the ownership of intellectual property and deemphasizes the "doing" comes to value ideas more highly than labor. Eventually the system creates a powerful new stratum of elite owners, far removed from the productive work of the economy. The rights of these new technological titleholders are legally sanctioned, but the legitimacy of their claims often remains dubious because of the debt they owe to innovations that have been made possible only by years or decades of collective advances.

It doesn't take much imagination to see the vast disparities of a system in which the production of every material thing, from clothing to computers, is relegated to a vast underclass of producers and manufacturers, while all the profit-enhancing innovations needed to compete in the global economy are tightly controlled by a wealthy cadre of technological titleholders. We see the trend already in agriculture, for instance, as agrochemical firms profit handsomely through the proprietary control of genetic information, leaving the modern farmer disenfranchised from the valuable information contained in the crops he or she sows, tends, and harvests.

The emerging system raises many troubling questions. Can an economy of ideas thrive in a free market built to handle the private ownership and dissemination of tangible commodities? Can our system for apportioning rights to knowledge assets allow individuals to reap rewards while also preserving the potential for rich future harvests from the knowledge fields? At the heart of these difficult questions — and at the heart of the new economy — lies our understanding of the role of human innovation: how innovation occurs and how it is related to the organically evolving store of knowledge.

To review both the rhetoric and the realities in this ill-understood area, I ask you to join me on a brief, eclectic journey through the landscape of invention in the twentieth century.

The Myth of the Knowledge Bazaar

Our tour begins at the National Inventors Expo in Washington. The annual exhibition, cosponsored by the U.S. Patent and Trademark Office, draws inventors from around the country to display — and hawk — their patented creations. With an atmosphere that falls somewhere between a high school science fair and a late night on the Home Shopping Network, the event resonates with a deep-seated tenet of American mythology: that entrepreneurial inventors built — and continue to drive — the country's economy.

As we enter, the exhibition hall is festooned with red, white, and blue bunting, a shiny electric car gleams near the entrance, and a myriad of gizmos is arrayed before us, each conjured by an individual inventor. Ronald Scott came to the expo from North Carolina with his patented hinged plastic case designed to enclose car batteries and minimize acid spills. Charles Popenoe shows off his patented Smartbolts, whose tops change color when sufficiently tightened. Twelve-year-old Akhil Rastogi demonstrates a patented spout he calls "E-Z Gallon" that allows kids to pour a gallon container of milk by tipping it rather than lifting it up. Septuagenarian Horace Knowles, at his booth, hands out examples of his patented bookmark, which lets you know by its orientation exactly where on the page you stopped reading.

These are the idealized inventors of our imagination. In the model system showcased here, good ideas are rewarded and individuals are spurred to innovate in their fields. Their discrete contributions enhance our lives and rarely overlap directly with one another. In this way the expo feels like the very essence of the free market itself, a kind of bustling "knowledge bazaar," where independent entrepreneurs vie for the attention of potential customers.

In case the connection between a gimmicky bookmark and technological progress seems strained, a prominent exhibit in the back corner of the hall commemorates the nation's canonized inventors — many of whom have huge U.S. companies and institutions named after them: Bell, Deere, Dow, Edison, Otis, Goodyear, Westinghouse. This corner exhibit is sponsored by the National Inventors Hall of Fame — an organization in Akron, Ohio, whose displays also adorn the lobbies of the U.S. Patent Office's complex in Crystal City. Their presence is not accidental. These venerated inventors are commemorated by the Patent Office like patron saints, inspiring the hopeful and reinforcing the agency's sense of mission.

Bruce Lehman, the commissioner of the U.S. Patent and Trademark Office, likes to invoke the memory of Thomas Edison. An unabashed enthusiast of the entrepreneurial spirit of the nation's inventors and a staunch advocate of an expansive system of intellectual property protection, Lehman notes, "When I walk into my office every morning, I see the patent model of Thomas Edison's light bulb sitting there, greeting me at the door." The display, he says, reminds him that "intellectual property protection, patents and copyrights have been a major part of the economic growth of America from the very beginning."

Lehman's contention is indisputably true. But today's system of intellectual property protection exists in an economy far different from the world of the Inventors Expo, in which ideas — manifest in clever new products — compete head to head in the free market. On the contrary, as we have seen, today's high-tech economy increasingly sanctions monopoly holdings over knowledge assets designed, at least in part, to exclude would-be competitors. In this terrain, ideas are more likely to compete head to head in the courtroom than in the marketplace. As business analyst Peter Drucker succinctly states, "The knowledge-based economy does not behave the way existing theory assumes an economy to behave." Quite simply, the ownership of knowledge seems to operate differently from the ownership of tangible goods, a fact that has been obscured because the industrial Second Wave packaged technological innovation into a parade of new consumer goods, from televisions to toasters. But today the emphasis is shifting from toasters to tollbooths: players are learning to cash in on new concepts by requiring royalty payments from anyone who wants to do business in a given area.

You can see the increasing power of these technological tollbooths throughout the high-tech economy. In software design, new players face a gauntlet of private claims on the building blocks they need to create new programs. And even making good-faith royalty payments doesn't diminish the very real threat that a software firm might unwittingly be infringing any one of the tens of thousands of software patents now working their way through the patent process. A similar thicket of private claims to seminal techniques pervades the field of biotechnology, where firms engaged in gene splicing have to pay hundreds of millions of dollars in royalties just to use the basic techniques essential to their business. One such biotechnology patent alone has earned Stanford University more than $150 million.

Clearly, the current model of innovation differs dramatically from that offered at the knowledge bazaar. How, then, did this model come about? We can find some answers by revisiting one of the most renowned fonts of technological innovation in modern history.

The Legacy of the Light Bulb

Thomas Edison's laboratory in West Orange, New Jersey, offers an extraordinary resource: a chance to glimpse a formative period in modern technology, during which the notion of commercially oriented research and development was forged. Built in 1887 in a drab, semi-industrial neighborhood forty minutes from Manhattan, Edison's laboratory fills two fenced-off blocks with a cluster of vaguely fortresslike brick buildings. The facility offers a wealth of insights into the origins of our views about intellectual property.

As almost everyone knows, Edison's development of a working incandescent light bulb paved the way for a now-omnipresent electric grid, much of which still bears his name. But the breadth of his inventions is less fully appreciated. Edison's astonishing record of 1,093 patents far outpaces that of all other individual inventors before or since. His invention of the phonograph would make possible the music recording industry, just as moving pictures, also his brainchild, would eventually put Hollywood on the map. Less well known are Edison's key advances in batteries or his invention of the microphone and the mimeograph, to name just a few of his creations. His portfolio even included a patent on poured concrete, part of a partially realized plan in which workers using his technique could purportedly build the structural shell of an entire middle-income house in just six hours.

But even given the monumental impact of his inventions on our technological world, Edison's greatest legacy may well be the way he encouraged us to think about technology. In particular, he drove home the anachronistic lesson that ideas are embodied in particular products. We have so heartily accepted his view that we have adopted Edison's most famous product, the light bulb, as our icon for an idea. And yet we have increasingly departed from this product-oriented model.

Edison called his laboratory an "invention factory." As we stroll through the place, the contrast between this conception and today's disembodied notion of invention is palpable. Laid out around a paved courtyard, the facility was one of the first full-scale R&D complexes in the world. Cashing in on the initial fame and financial success he achieved with the light bulb, Edison created what he called the "best equipped and largest Laboratory extant." It boasts a metal foundry, two large machine shops, a fully stocked chemistry lab, a darkroom, and sound recording and film studios. Edison wanted to be able to "build anything from a lady's watch to a locomotive," an effort, he noted, that required the lab to carry "a stock of almost every conceivable material."

Today the vast storeroom in the lab's main building is testament to the central role materials played in Edison's R&D enterprise. Banks of small wooden drawers line the walls, each holding samples of different materials; scores of metal sheets, rods, and pipes are stacked neatly to the side. Standing before the jumbled backdrop, our guide, Maryanne Gerbauckas, superintendent of the Edison National Historic Site, recounts Edison's famous quip that the storehouse contained "everything from an elephant's hide to the eyeballs of a United States Senator." But a newspaper report from 1887 offers a more complete accounting: "eight thousand kinds of chemicals, every kind of screw made, every size of needle, every kind of cord or wire, hair of humans, horses, hogs, cows, rabbits goats, minx, camels . . . silk in every texture, cocoons, various kinds of hoofs, sharks' teeth, deer horns, tortoise shell . . . cork, resin, varnish and oil, ostrich feathers, a peacock's tail, jet, amber, rubber, all ores, [and] metals."

Edison put the exotic materials to use with surprising regularity. His notebooks, now kept in a cloistered vault on the site, show that he and his assistants tried no fewer than three thousand separate materials in their quest for an effective light bulb filament, testing everything from platinum to Japanese bamboo before settling on carbonized cotton thread. After much similar trial and error, Edison employed compressed rain-forest nuts to make the needle used in some of his earliest phonograph models.

Edison's tireless focus on translating his far-ranging ideas into products is tangibly evident throughout the West Orange facility, but perhaps nowhere as clearly as in a rarely viewed attic closet. Past rows of shelves holding uncataloged artifacts on the main building's top floor, Gerbauckas opens the storageroom door to reveal a staggering display: scores of phonograph horns — the component that amplified the early record player's sound — in every size and shape. Some of the horns are rounded, others angular; some are short and squat, others elongated, standing no less than six feet tall.

The contents of this closet illustrates the primacy Edison placed on the role of development in the making of products. It also underscores a fundamental distinction between Edison's era and our own. The rogues' gallery of rejected prototypes testifies to Edison's tenacity and offers a key to his success: trying out every design he could conceive of — a practice that lent credence to his famous maxim that "genius is 1 percent inspiration and 99 percent perspiration."

By equating ideas and products, Edison's invention strategy effectively fused research and development in a seamless process. When he invented a phonograph, he not only perfected a prototype, he built factories employing thousands of workers to make the machines. Edison even personally oversaw the construction and operation of a sound studio and the selection of many of the earliest musical recordings. In so doing, he finessed some of the difficult questions raised by his dependence on technological innovation. From the government's perspective, the risks Edison took in the expensive development phase, and the rich rewards the work delivered in innovative new products, seemed well worth the protection of the patents it awarded him.

Perhaps most significantly, in all his endeavors, Edison succeeded and failed at the hand of the market. Historian Gregory Field describes the phenomenon as Edison's insistence on "always tying the 'R' with the 'D.'" As Field puts it, he persisted in the view that "invention involves not just research, but research, development and marketing." Edison relied on his ability to keep his technology ahead of the competition. Though his phonographs offered the best sound reproduction of the day, he ultimately lost out in the market because he couldn't produce the machines as inexpensively as some competitors.

The Specter of Consolidation

The irony of revisiting Edison's invention factory is that in some respects it is so familiar. We have fully retained a Second Wave, factory-oriented view of ideas as products even as our economy has dramatically deviated from this model. We continue to conceive of ideas as tangible commodities even though, as we have repeatedly seen, they function increasingly as tollbooths or even barricades imbedded within the system and divorced from the marketplace.

The irony of Edison's workplace is augmented by its historical context. In many ways, his era mirrors our own. Edison came of age at a time when the promise and excess of industrialization were felt by all. Then, as now, a frontier mentality pervaded. A new set of moguls, many of them Edison's friends and acquaintances, was amassing huge fortunes. He vacationed with Henry Ford and Charles Goodyear; the Carnegies and Rockefellers were his contemporaries; and J. P. Morgan, the architect of the steel monopoly, was one of Edison's original financial backers in his development of the light bulb.

At the same time, that era was marked by growing concern with the stranglehold capitalists were gaining over the nation's infrastructure. Standard Oil's "trust," or cartel, was created in 1879, just months before Edison won his patent on the light bulb. In an economy that ran on oil, steel, and railroads, people soon realized that price fixing and collusion in those industries had undermined fair competition. These concerns ultimately led to the passage of the Sherman Antitrust Act of 1890 and further legislation to strengthen the government's power to break up monopolies in the two decades that followed. As Peter Lynch and John Rothchild recount, "When the twentieth century rolled around, it was obvious that something was wrong with the way capitalism was going. It had started out as a free-for-all when anybody with a good idea had a chance to succeed. It was turning into a rigged game dominated by a few giant businesses."

A hundred years later once again, something is wrong in the capitalist system, and it is likely to require a similar campaign to address it. While the problems in Edison's day arose from overt manipulation of the free market, today's problems are more insidious than price fixing, collusion, and unscrupulous business practices. Like the infrastructure of Edison's day, our collective know-how — represented in blueprints, codes, and formulas — makes up a kind of "infostructure," a framework of information and knowledge that allows our economy and our culture to function smoothly. The problems we face now stem not from a perversion of the market but from the inapplicability of our traditional conception of the marketplace to the vibrant new fuel at the heart of today's economy.

The end of the Cold War has prompted near-global unanimity that free market competition is unrivaled in its ability to efficiently distribute goods. But we are coming to realize that this efficiency does not translate into the conceptual realm. Too often the exclusive ownership of ideas allows private titleholders to shovel perilously near the taproot of our collective heritage of knowledge. Private claims to disembodied knowledge assets are increasingly seen as a threat to broad public interests even as they fill individual coffers.

Despite this disturbing picture, regulators and titleholders alike determinedly treat the ownership of knowledge just like the ownership of any other tangible commodity, attempting to extend the rules that have served them well in one broad domain into a very different one. These groups confidently argue that any particular problem that arises requires merely a technical or semantic fix — a new footnote added to the regulatory rulebook. From their perspective, the unfettered free market has worked so successfully for producing and disseminating tangible goods that it has become a kind of inviolate dogma. Any intervention, no matter how clearly merited for the continued viability of the system, is to be vociferously resisted.

When doctors began in earnest to patent medical procedures, for instance, these groups resisted the conclusion that this category of formerly shared knowledge should be off limits for private claims, as it is in most other industrialized countries. They lobbied instead for a legalistic exception that restrains medical practitioners from suing one another while otherwise keeping the system intact. When the merger to form the vast pharmaceutical firm Novartis created a multinational company with broad monopoly patents over human gene therapy, a similar contingent of advocates for the status quo resisted a systematic review of how such broad claims affect progress in the field. Instead they allowed a single dominant corporation its powerful patent portfolio but, in this one instance, suggested an exception requiring the company to license its human gene therapy patents to all competitors.

This ad hoc, incremental approach may work for now. But it will never keep pace with the burgeoning procession of expansive private ownership claims on the horizon. What is needed are guidelines about when the overriding public interest requires restrictions on private capture of formerly shared knowledge assets. Unless we tackle the issue head on, proliferating private claims will choke productivity, magnify current inequities, and erode our democratic institutions.

The most troubling piece of the equation may be the threat of a new kind of consolidation. The paradox is this: the wealth-generating engine of the new economy is innovation, which is a decentralized, grassroots, individualistic endeavor. You don't need to build a large factory or amass a stockpile of raw materials to solve a key engineering problem or think of an important new idea. And yet the notion of a free, competitive "marketplace of ideas" seems, if anything, to be receding in the new economy. Instead, we see an unparalleled trend toward consolidation. In the thrall of technological development, and the grassroots potential of innovation, however, many overlook the power of the emerging "fiefdoms" of knowledge assets, portraying an idealized vision that has more to do with the Inventors Expo than today's global economy.

In a document immodestly titled "A Magna Carta for the Knowledge Age," for instance, authors Esther Dyson, George Gilder, Jay Keyworth, and Alvin Toffler make much of the Third Wave economy's ability to "de-massify" society, offering products specifically tailored to individual needs rather than mass-produced. These authors, enamored of the possibilities offered by the decentralized nature of the Internet as opposed to, say, broadcast television, cast a blind eye to the realities of monopoly ownership.

In their view, "Second-wave ideologues routinely lament the breakup of mass society. Rather than seeing this enriched diversity as an opportunity for human development, they attack it as fragmentation and balkanization. But to reconstitute democracy in Third Wave terms, we need to jettison the frightening but false assumption that more diversity automatically brings more tension and conflict in society."

Unfortunately, these high-tech gurus are wrong on all counts. First, as we've seen, from the ivory tower to the Ecuadorian parliament, the tensions and conflicts of the knowledge-based economy are not a "false assumption" but a bona fide and growing threat. We are constantly barraged with intractable patent disputes and high-profile lawsuits over intellectual property claims.

Equally compelling is the documented trend toward a stronger kind of monopoly ownership than ever before. If we need to jettison something, it is the vision of a decentralized, self-correcting, free-market knowledge bazaar. Consider, for instance, that $1 trillion in corporate mergers took place in 1997, a pace of mergers unrivaled, according to the *New York Times,* "since a wave of industrial takeovers created the great oil, steel and auto companies at the beginning of the century." Between 1994 and 1997, according to this account, some 27,600 companies have merged, more than in the entire decade of the 1980s, which was previously considered the high point for mergers.

Can the widely respected authors of the new "Magna Carta" truly believe that the new economy is "de-massifying" in the face of such statistics? Today's companies may be able to more readily personalize product lines, but that superficial variety belies the overarching trend toward consolidation. In many high-tech sectors we are coming to recognize the dominance of fewer, larger companies as a hallmark of the knowledge-based economy.

In some sectors consolidation already approaches the limit. The U.S. defense industry, for instance, which in the early 1990s comprised more than fifty military contractors, is now dominated by three massive firms: Boeing, Raytheon, and Lockheed Martin. Lockheed Martin alone has acquired twenty-six different companies in the past several years. Eight conglomerates, including Time Warner, Hearst, and Viacom, now overwhelmingly dominate the converging media and publishing sectors. As we have seen, ten firms control 80 percent of the agrochemical market. Similar consolidation is currently under way in an array of sectors, from pharmaceuticals (where twenty firms now control 57 percent of the global drug market) to the global telecommunications industry, where behemoths like Bell Atlantic, MCI, and Japan's NKK predominate. The trend even holds in newer sectors like the software industry, where, despite the existence of many hopeful start-up firms, Microsoft's dominance is almost unprecedented in the annals of business.

These striking figures suggest that today's economy is not headed in a decentralized, "de-massified" direction at all but instead has entered a period of unprecedented consolidation.

10

The New Monopolies

TO UNDERSTAND how an economy based on the ownership of ideas functions, it is helpful to consider the case of Jerome Lemelson, the most prolific individual inventor of modern times. Some fabulously wealthy people have made fortunes owning oil wells, vineyards, or railroad lines. Jerome Lemelson owned ideas. With more than five hundred patents, Lemelson was a seemingly limitless font of ingenious ideas. His accomplishments rank him, in fact, as second only to Thomas Edison in number of U.S. patents. Yet most of Lemelson's inventions are so conceptual that he would be hard pressed to bring any of them to display at the annual Inventors Expo.

In 1995, two years before his death at age seventy-four, we find Lemelson at his grand home office overlooking the mountainous northern shore of Lake Tahoe. After years of hard work and struggle, Lemelson has arrived at an impressive pinnacle of influence and wealth. Gazing out from his breakfast nook at the magnificent alpine lake, Lemelson regards a view as pristine and intangible as the empire of knowledge assets he presides over: a multimillion-dollar enterprise that produces few products and employs fewer workers.

Lemelson never ran a renowned entrepreneurial laboratory such as the ones Edison or Alexander Graham Bell once oversaw. He relied instead on his imagination and his omnipresent handwritten notebooks of ideas. Nor did his inventions spawn a vast corporate empire bearing his name, as did Second Wave inventors like John Deere, Herbert Henry Dow, or George Westinghouse. And yet his astonishing portfolio of patents has undeniably touched all our lives.

Lemelson patented the "recognition tones" that allow fax machines to reliably communicate with each other, earning him royalties from virtually every fax machine manufacturer in the world. He staked a claim on a component of the bar-code scanning machines that are now nearly ubiquitous at supermarket checkout counters. And he patented the machine vision used by robots on many of the world's assembly lines. Lemelson filed for patents on the idea of a video camcorder in the mid-1970s — almost a decade before the product appeared on the market. He even patented the concept of the tape-drive mechanism that engineers ultimately built into the Sony Walkman — a development that won him a lucrative licensing arrangement with Sony and 105 other companies.

The list of concepts Lemelson claimed rights to is as broad as it is impressive. His portfolio includes claims on everything from illuminated highway markers to medical procedures for treating brain cancer. But it is in the conceptual nature of his work that Lemelson most distinctly mirrors his time. Perhaps his greatest genius was in understanding how to capitalize on our system of parceling out ownership rights to intangible ideas. The man indisputably conceived of — and laid legal claim to — key technological advances that contributed to many of today's high-tech consumer goods. But Lemelson himself rarely translated his ideas into actual products. The embryonic quality of his ideas — divorced from particular products — was one of his hallmarks. It also marks a problem inherent in the knowledge-based economy.

Charging Rent on Dreams

During Lemelson's lifetime, his public relations representative, Anne Cunniff, of Lipman Hearne, regularly defended his conceptual claims and lack of workable prototypes by explaining that he was "a thinker, not a tinkerer." The explanation sounds facile, but it is more useful than it would first appear. To be sure, much of the credit for Lemelson's success does stem from his role as a thinker: his uncanny talent for envisioning connections and new applications most of us would never dream of. Seeing Velcro on his wife's belt for the first time in the 1950s, for instance, Lemelson quickly imagined — and patented — the popular game in which Velcro-covered Ping-Pong balls are thrown at a target.

But Lemelson's role as a "thinker" is also at the heart of the most devastating critiques leveled against him. Critics often angrily charge that he manipulated the patent system to squeeze money out of big companies. He did construct a working prototype of the Velcro dart game, for instance, but detractors charge that Lemelson never made — or even figured out precisely how to make —

most of his lucrative inventions. His widely accepted claim to having "invented" the camcorder, for example, consists essentially of the realization that a hand-held video camera might contain a videotape cassette on which to record pictures and sound. It can be argued that it was a novel conception. But even Lemelson makes no claim to have actually constructed such an item in the 1970s, only to have been the first to take the notion to the Patent Office. What made his claim viable is that videotape already existed, as did portable sound recording equipment, whose design Lemelson proposed to appropriate. In such cases, the conception alone is arguably "enabling" enough, in the lingo of the Patent Office, to merit a patent.

How much credit — and recompense — does Lemelson deserve for his often embryonic, conceptual efforts? This is not an idle question but one that embroiled him in dozens of multimillion-dollar lawsuits. As an example, let's consider a notably vague and broad patent Lemelson received on automated sensing devices, one of a group of inventions covering machine vision that Lemelson called his "greatest accomplishment as an inventor." Here's the abstract of U.S. Patent No. 5,351,078 as it was issued in 1994, entitled "Apparatus and Methods for Automated Observation of Objects":

> Apparatus and methods are disclosed for automatically inspecting two- or three-dimensional objects or subjects. A detector and the object are moved relative to each other. In one form, a detector, such as a camera or radiation receiver, moves around an object, which is supported to be rotatable such that the detector may receive electromagnetic energy signals from the object from a variety of angles. The energy may be directed as a beam at and reflected from the object, as for visible light, or passed through the object, as for x-ray radiation. Alternatively, the detector passively receives energy from the object, as in an infrared detector. The detector generates analog image signals resulting from the detected radiation, and an electronic computer processes and analyzes the analog signals and generates digital codes, which may be stored or employed to control a display.

As even a casual reader can see, the patent construes Lemelson's notion of "automated detection" as broadly and generally as possible. Patent attorneys describe patents like this one as "ink-blot tests," rightly noting that you can read almost anything into the claims.

The power of a broad patent claim can also be significantly increased when the patenting process takes a long time. It is hard to believe, but Lemelson filed his initial claim on automated detection in 1954. For no less than forty years Lemelson's lawyers engaged the U.S. Patent Office in an iterative process of "divisions" and "continuations" of the original claim, during which time Lemelson was legally allowed to modify the details to incorporate mention of ongoing technological developments in the field.

This technique is a classic example of what some call a "submarine patent." In this scenario a broad and vague patent application is filed very early in the development of a new technology. Over a lengthy approval process, the claim is honed and fleshed out. Years later, when the patent is finally granted, the early, seminal claim "surfaces" — bolstered by the backing of a government-sanctioned monopoly — to wreak havoc on an industry that has already brought to market the items covered in the patent claim.

In Lemelson's case the notion of automated detectors was surely farsighted when he first filed it. But four decades later, when the patent was issued, automated detectors were already widely used in a variety of applications from manufacturing to airport security. Once his patent was belatedly issued, Lemelson's technique — one he repeated throughout his career — was to demand royalties from the companies with existing products that could be construed as infringing his broad claims. The odds for success were in his favor. He could win either by receiving royalties from the blindsided companies or by suing the holdout firms, building a strong legal case that relied, at least in part, on the early initial filing date of his patent claim.

"If you file often enough and long enough, you can eventually get a patent on almost anything," explains patent attorney Robert P. Bell, who has studied Lemelson's legal tactics. "Given sufficient time and money, I could probably get you a patent on the wheel." All it takes, he adds, is to "wear down the examiner for twenty years or so." Lest Bell's boast seems exaggerated, it is worth noting that since he made that comment, someone actually has accomplished the feat. In January 1998, U.S. Patent No. 5,707,114 entitled "Vehicle Wheel," was issued to a Connecticut inventor. The patent's abstract claims exclusive rights to the wheel, including "an annular rim, a central hub and a plurality of spoke portions running between the rim and hub."

Lemelson's critics emphasize the enormous gap between describing an invention in words and building something that works. As the adage goes, "The devil is in the details." U.S. businesses spend far more on the development part of the R&D equation than on research alone. In fact, a significant rationale for the patent system is a tacit agreement between the government and the inventor: the government offers a limited monopoly with the expectation that the protection will spur a new product to the market. Absent this painstaking development component, critics say, Lemelson's contributions are of questionable worth.

The problem of disembodied, conceptual patent claims prompted the British journal *The Economist* to suggest in 1994 that the United States overhaul its entire patent system. As the editors argued, "True, a few inventors may watch in outrage as their ideas run away from them. But a little more insecurity for inventors may be no bad thing. Doing things with ideas is what makes innovation a reality, not charging rents on dreams. The doing merits the reward."

Up the Ladder of Abstraction

If the present system disproportionately rewards ideas rather than their applications, it is also plagued by a related problem: the difficulty inherent in distinguishing one idea from another. As Lemelson's patents clearly indicate, largely disembodied concepts like automated detection can be slippery indeed. It is almost impossible to delineate clear, rational boundaries between one idea and another. To a large extent, Lemelson's success rested on his talent for couching his insights and innovations in the broadest possible terms. Clearly, the system rewards those patent claimants who can move up the ladder of abstraction to stake their claims as generally as possible. In this way, deft idea magnates can carve out the biggest swath of protected, conceptual turf.

The drive to portray a conceptual patent as broadly as possible is explained clearly by William Budinger, CEO of Rodel Inc., who holds more than three dozen patents. As Budinger explains, "I could file a patent on a red chair, if one had never been made. But somebody else before could have filed a patent on a chair. If both patents issue," he says, "I cannot make my red chair without the permission of the guy who makes the chairs." Clearly, in such a system, any inventor of a red chair would be well served by applying to the patent office for the broader "chair" patent.

We have seen precisely this phenomenon in cases such as the Agracetus soybean patent, in which the company argued that its single genetic manipulation of a soybean cell entitled it to a patent covering all genetic engineering of soybeans by any means. The phenomenon was also on display in the claims of the software firm E-data that its program allowing secure financial transactions on the Internet affords it the broad right to demand royalties from any other company working to facilitate commerce in cyberspace.

Budinger's comments point to a related problem. The courts might be able to fairly distinguish between particular types of chairs. A jury might be able to rule effectively whether someone's maroon chair infringes someone else's red-chair patent. But the task becomes markedly more difficult at higher levels of abstraction. For example, my new benchlike invention shares some features of a chair. Does this mean it infringes your broad "chair" patent? It can be exceedingly difficult — if not impossible — to fairly and consistently adjudicate such cases. Budinger's caricature describes a system that rewards those who act fast to make a broad claim on conceptual turf. And, like the children's game of musical chairs, the net result is one lucky winner — in this case an inventor who can demand royalties from everyone else who wants to sit down.

The unproductive and unfair nature of such a system is obvious. But its viability is also called into question given the conundrum of where ideas spring from in the first place. What ownership rights, if any, should my specific, innovative chair design earn for me? The critic Northrop Frye captured this dimension of the problem in 1957, when he noted famously that "poetry can only be made out of other poems; novels out of other novels." Bruce Hartford of the National Writers Union underscored the point emphatically in 1997 to an audience of computer professionals by observing, "All new knowledge, every single piece of new intellectual property, is built on the intellectual foundation of what has gone before. You cannot be a writer, or any other kind of creator, without also being a reader and a researcher."

The result is a situation that is nearly impossible for the legal system to sort out. As the patent system allows inventors to move continually "upstream" along the river of generality in the concepts they lay claim to, we can fully expect to see a surge of broad, contested, overlapping claims to valuable pieces of the knowledge economy. And alas, this explosion of litigation has already begun in earnest.

Courting Disappointment

Lemelson's tale is a story of a personal fortune built upon broad, conceptual ownership claims. The results — multiple, protracted multimillion-dollar lawsuits — offers a fitting allegory for the emerging economy as a whole. To be sure, legal battles — even those over intellectual property — are nothing new.

But their prevalence, complexity, and increasingly intractable nature are unprecedented.

Here again, Lemelson reflects his times: his early experiences forged a litigious streak that remains legendary in some technological circles. In his early days as an inventor of children's toys, Lemelson attempted to peddle his ideas to toy manufacturers, only to see them copied by others. After years of believing that his ideas were being stolen, he vowed to take up the cause of independent inventors everywhere, fiercely defending his intellectual property claims. In the years that followed he came to view himself as a champion of the plight of the independent inventor — a lone David standing up to the Goliaths of the corporate world.

Lemelson's multimillion-dollar lawsuits, led by a now-famous law firm in Nevada, won him huge settlements from some of the world's largest corporations. In one set of cases in the early 1990s, he won more than $500 million from auto manufacturers and other firms for their widespread use of automated manufacturing systems that drew upon some of the machine vision patents he had initially filed decades earlier. However, as Lemelson confided with some regret toward the end of his career, his proliferating lawsuits required him to spend a third of his prodigious annual earnings and even more of his time furthering his legal cases.

Despite all that time, effort, and money, though, Lemelson was forced to confront the inescapable fact that, in the hands of the legal system, even the most airtight patent claim can deflate into a worthless pile of paper. In a classic example from relatively early in his career, Lemelson won $71 million from the Mattel Company in a bitterly fought battle in which he claimed ownership of the idea behind Mattel's Hot Wheels cars and tracks. But a subsequent verdict in the court of appeals (the nation's highest court for routine patent cases) ultimately overturned the award, upholding Mattel's claim to the toy.

Overall figures are scarce, but Lemelson's experience is not an anomaly. Patent lawyers routinely report an expanding docket of intellectual property battles. Intellectual property law, widely hailed as a growth area for attorneys entering the profession, already draws more young lawyers than ever before. Most of these cases are settled before reaching the courts, but in those few areas where analysts have tabulated the cases, the figures are dramatic. One study documented that patent litigation in the field of biotechnology increased 69 percent between 1995 and 1996 alone.

But the full story is told not in the sheer quantity of intellectual property disputes but in their scope. In 1997, for instance, Digital Equipment Corporation (DEC), one of the most venerable computer firms, sued Intel, the world's largest computer chip maker, over rights to the $20 billion empire of Pentium chips Intel has sold to power the latest generation of computers. DEC claimed Intel stole its techniques to make the Pentium chips. Intel claimed it didn't. Finally, as part of a merger deal (with another computer giant, Compaq, gobbling up DEC), Intel purchased DEC's portfolio of relevant technology for $70 million. Who knows what to make of such an outcome? Did the Pentium chip owe a debt to DEC? Of course it did, just as it did to decades of publicly funded basic research in computer science. But what kind of compensation is due for building on existing knowledge? It is a deeply vexing question.

While DEC and Intel battled over who would own the world's most valuable hardware, Sun Microsystems took Microsoft to court, alleging illegal use of Sun's Java programming language. And in biotechnology, Hoffman-La Roche, a Swiss-based firm, is embroiled in a lengthy patent dispute over what may be the field's single most valuable technology — a technique called PCR, or polymerase chain reaction, that makes most modern gene sequencing possible. In 1995 Hoffman-La Roche, which had acquired proprietary rights to PCR technology several years earlier, even drew forty of the world's top research universities into the fight. The company considered the technology so valuable it demanded royalties even from academic researchers who claimed to be using it solely for noncommercial purposes.

Intractable disputes like these represent the Achilles' heel of the knowledge economy. By presuming that the new regime can run on the same rules of private property that fueled the industrial age, we have created a tinderbox. Unlike land or other forms of tangible property, knowledge is slippery, interwoven, and often needs to be shared to be of any use. It cannot easily be cordoned off and parceled into discrete packages. Unfortunately, however, this is what is all too frequently being attempted.

Barricading Technology

High-stakes battles over intellectual property do more than sap untold millions from the economy. The fallout from these cases goes far beyond lost work hours and internecine corporate bloodletting. Patent infringement cases are among the most costly types of litigation in the U.S. legal system. Even the average, run-of-the-mill patent infringement case that goes to trial now costs litigants $1.2 million in legal fees, according to the Virginia-based American Intellectual Property Law Association. As Greg Aharonian notes, even a preliminary court skirmish on a patent issue can cost a company $100,000 — an amount, he points out, that might be an entire year's profit for a small start-up firm.

Given the momentous stakes and high costs involved, the mere threat of a patent suit is often enough to deter competition, especially by smaller, innovative new firms. Many large companies have already developed "defensive patent portfolios," using patents as bargaining chips to protect themselves against litigation and to ensure prospects for cross-licensing. As one commentator describes it, "In the struggle to compete, protecting intellectual property has become as important a weapon as creating it."

Take the case of Dow Chemical Company, which in 1996 chartered an in-house "intellectual-asset management" group to oversee its portfolio of 30,000-plus patents. By more aggressively charging other firms for using Dow's intellectual property assets, Gordon Petrash, who heads the group, says Dow will enhance its "licensing value stream" from $25 million in 1994 to a projected $125 million in the year 2000. But "the real bottom line," according to Petrash, will come in the future, as Dow more actively pursues control of promising new intellectual property assets, a technique of "value creation" he estimates will ultimately be worth "billions and billions."

There is no dearth of enterprising companies offering services for this new business climate. Consider, for example, SmartPatents, Incorporated, an up-and-coming firm that has partnered with the giant accounting firm Price Waterhouse to help other companies build and protect their intellectual property portfolios. SmartPatents' CEO, Kevin Rivette, calls patents "the new global currency of technology" and explains that businesses need his firm's services because intellectual property is commanding an increasing portion of what he calls "a company's overall valuation equation." As SmartPatents' literature explains, the firm can warn a business when competitors' ownership claims are "surrounding its core technologies" and can help firms "develop counter-strategies" to protect and defend their assets.

Owning the "Infostructure"

Clearly, today's successful corporations are developing a strong arsenal of strategies to adapt to the realities of the knowledge-based economy. More often than not, the result is a consolidation of techniques and know-how within a given field, a result fostered by the present patent system. But consolidation often proves to be a deeper, endemic problem when individual players manage, through whatever means, to gain control of powerful concepts.

A clear example of the problem can be seen in the intriguing antitrust case the U.S. government is currently building against Microsoft. At the heart of the legal case is the fact that Microsoft's operating system, Windows, controls 86 percent of the market. By running this software alone, nearly nine out of every ten computers in operation make use of Microsoft's intellectual property. At issue is the way the company has leveraged this enormous base of customers to build a similar kind of dominance in related areas. In 1994 the U.S. government charged that Microsoft, by forcing computer manufacturers to include its Internet Explorer (a piece of software used for viewing and interfacing with the World Wide Web) with every copy of Windows 95, had unlawfully "tied" the product to the purchase of its operating system. A U.S. district court issued an injunction in December 1997 barring Microsoft from forcing computer makers to accept Internet Explorer as a bundled part of Windows. Since then, the mounting high-stakes battle has included federal hearings and legal appeals. Many of the states' attorneys general have joined the fray. The final outcome of the case will doubtless be years in the making.

Reviewing the charges, some analysts, like Robert Kuttner, have argued that there is no fundamental difference between the government's antitrust case against Microsoft and previous fights against classic anticompetitive behavior. But this argument tells only part of the story. Underlying the government's case against Microsoft is a new economic view about high-technology markets that merits close attention.

The predominant view of mainstream economists in the 1960s and '70s held that free markets work best with a minimum of governmental intervention. Today, however, a growing number of both economists and legal specialists believe that in the knowledge-based economy, markets don't necessarily self-correct but rather require intervention to protect the public's overriding interests.

Perfect markets work, as nature does, with negative feedbacks. When an ecological balance tips, as when a deer population exceeds a given region's carrying capacity, a corrective feedback — namely, starvation — occurs that reduces the number of deer. In this classic scenario, only the strongest individuals survive and equilibrium is restored, benefiting survivors.

In the Microsoft case, however, the government's antitrust lawyers are persuaded that what economists now describe as "network effects" have given Microsoft an unfair advantage. As the argument goes, in situations — prevalent in the conceptual realm — where it is desirable for goods to be compatible with one another, the value of a product increases with the number of people who are using it. In other words, the more people using a computer operating system, say, the more desirable and useful it becomes to others. Ultimately, network effects tend to lock in standards such as VHS videotape, CD-ROM formats, and Internet software protocols. When a lucky firm like Microsoft gains exclusive rights over such a technological standard, it has the opportunity for an extremely powerful new type of monopoly.

Network effects offer a classic example of a positive-feedback system. Unlike the market system, positive-feedback arrangements have no built-in methods for self-correction; they tend toward disequilibrium. Precisely these effects, the U.S. government contends, have caused Microsoft to have an unfair advantage in the marketplace.

Recognizing the systemic problem presented by network effects, many observers are calling for government action. As Daniel Rubinfeld, chief economist in the Justice Department's division prosecuting the Microsoft case, explains, "There is still an honest debate about exactly what role government ought to play, and people are going to differ, but there are very few economists I have talked to who would argue that leaving it to the market is always the best solution. We are just not in that world anymore."

Our tour of inventors past and present raises grave doubts about our faith in the dogma of a self-correcting free market for ideas. We are gradually waking up to the fact that knowledge assets are inescapably part of a shared infostructure. The trick will be to find a way to allow inventors and innovators to freely build upon the strong foundation of past discovery and insight without eroding the public domain. This will not occur by the market alone but through corrective, democratic action by the public looking out for its pooled interest.

Another way to look at the Microsoft antitrust case, for instance, is that one company has gained exclusive private control of a language many people wish to speak. What the firm owns can thus be seen as a valuable piece of our infostructure. The fact is, whether the control is de jure, via a government patent, as in Lemelson's case, or de facto, via network effects, exclusive private ownership of the infostructure is inherently problematic. This is true whether we speak of the genetic or medical information drawn from the biological systems of the natural world or the storehouse of human knowledge handed down from one generation to another. These assets will wither unless they are protected, promoted, and preserved as a shared heritage.

II

The Most Precious Asset

THE TALES OFFERED in this book are the classic skirmishes of the frontier. Excitement and opportunity go hand in hand with a messy, greedy lawlessness. Almost everyone involved acknowledges the need to establish equitable new rules of the road, even as they actively vie to stake their own private claims.

Amid all the chaotic action, however, the overarching challenge is clear: we can use our prodigious knowledge tools to improve public access to every kind of information, to increase participation, reinvigorate education, and build robust new systems of commerce and governance that will foster a renaissance of new discoveries. Or we can let parochial private interests shape the future for us.

The first task is to recognize that a confluence of forces is forging a new global economy based on the private capture of knowledge. This new regime is being implemented swiftly across many disparate high-tech fields, supported and codified in an array of legal, political, and international actions. The old system of commerce has been stretched to the breaking point in an effort to accommodate the privatization of concepts, information, blueprints, formulas, and even life forms. Yet, a growing body of evidence shows how differently these intangible knowledge assets operate in the economy compared to the classical goods and services it was built to disseminate.

As we have seen, some players in the new economy are already experiencing the profound and vexing problems that emerge when knowledge is treated exclusively as a commodity. But the brunt of the ill effects will be felt in years to come. In fields from medicine to software, participants will learn the hard way that dividing cutting-edge knowledge into private parcels hampers the flow of information and stifles innovation. Even more troublesome in the near term, entrepreneurs will become increasingly tangled in bitter and costly battles over rights to these slippery, intangible knowledge assets. Finally, those who seek to capitalize on knowledge derived from publicly funded research or from the fertile public domain will face growing public opposition. Already the public's appreciation of our shared stake in these powerful assets is undermining the rhetoric of unfettered private ownership as the exclusive ownership of ideas and concepts is increasingly seen to clash with democratic principles of equal access and equal opportunity.

In the years to come, the most hotly contested battles are likely to occur in two areas that will powerfully shape the development of knowledge in the next century: the quest to understand the human genome and the collection of networked computers we have come to call cyberspace.

A Shared Genome

The multibillion-dollar international effort to map the human genome presents an unprecedented opportunity. Decoding the makeup, and ultimately the functioning, of the roughly one hundred thousand genes that direct cells to build human beings promises to revolutionize biomedical science, leading to new drugs and therapies and a deepened understanding of disease and aging. But mapping the human genome also puts our system of ownership of knowledge assets to one of its most difficult tests: should researchers be allowed to own particular segments of the human genome they decode?

Scientists hope to have a complete map of the genome within the first decade of the new millennium. Already some of the most tantalizing pieces have been mapped. The job of decoding, or sequencing, this genetic information is enormous. Each gene — a fragment of the basic building block of life called DNA — contains somewhere between 150 and 3,000 base pairs of nucleotides, expressed by their initials as a string of the letters G, C, A, and T in an intricate pattern that encodes the information. And as if that were not a daunting enough code to crack, the so-called working genes make up only roughly 3 percent of the human cell's inheritance. The rest of the genome — some 97 percent that needs to be sifted through — is noncoding DNA, whose seemingly random repeating patterns have no known function.

By mapping the code that directs human cells, scientists are taking the first step toward the goal of discovering how each piece works with the others to determine human traits and characteristics. The complete genome map —

which will be equivalent to some two hundred volumes each the size of a 1,000-page telephone book — represents an essential tool to begin this research. But already private companies and academic research groups are staking exclusive claims to this vital information.

Today, when teams of researchers "tag" a gene by identifying the initial string of letters that can be useful in marking its spot on the genome, they often apply for a patent, even though they usually don't know the gene's function or role. The U.S. Patent Office now says it has applications pending for some five hundred thousand tags — known as Expression-Sequence Tags, or ESTs. After years of equivocating, the Patent Office remains unsure whether to grant patents on this basic, preliminary information from the Human Genome Project.

But the debate over ESTs is merely one of many debates over private rights to genomic information that are already dividing the research community. Even many scientists who support the notion of patenting useful products derived from the human genome, for instance, denounce the patenting of ESTs as devastating to the development of the field.

This debate is too important to leave to a rarefied coterie of scientists and patent lawyers. Without public intervention, we risk a scenario in which large corporations will own the rights to the genetic legacy of our species. In this arrangement, we will needlessly cede control of the development of a technology that intimately affects the health and future development of our species to private feifdoms. In fact, the trend is already under way to a startling extent.

Many scientists argue that private ownership provides the needed incentive for scientists to accomplish the daunting job of sequencing the genome in a timely fashion. Plus, they emphasize, patents require public disclosure, far preferable to secret private claims.

As George Poste, a researcher at SmithKline Beecham, which has a stake in one of the largest private databases of decoded human genes, says, "Superficial slogans, such as 'genes are merely research tools' and 'no patents on life,' will erode the foundation of intellectual property rights needed to convert innovative research into new products, to the future detriment of medicine and human welfare."

But many other leading biologists bristle at that idea. "Contrary to the claims of the biotechnology industry, gene patents retard progress in the biomedical arena, introduce secrecy where openness is essential, and slow the publication and sharing of important results," MIT biologist Jonathan King argues, for instance. King emphasizes that genetic engineering techniques have been developed through some four decades of public funding in biomedical research. As he puts it, "Venture capitalists and pharmaceutical firms played no significant role in these historic breakthroughs." But over the past decade "private corporations have moved to appropriate this knowledge."

The largest repositories of human genetic sequence data are now believed to reside in private hands. According to one estimate, the U.S. Patent and Trademark Office has already issued some eight thousand patents on human genes, or methods and techniques directly related to them. The battle now centers on what kinds of genetic information can be privately owned.

A key issue is the need for open access to this information. Most information about human genes is currently being shared. At the end of 1995, one scientific team made headlines by placing on the Internet the lengthy sequence for a gene believed to be linked to breast cancer — a potentially lucrative discovery. The scientists at Washington University in St. Louis and the Sanger Center in Britain said they wanted to show that such data best benefits society when it is made available freely and quickly. Since that time a wealth of genome-related information has been immediately released on the Internet. It has been estimated, however, that public access to at least 15 percent of the information compiled so far is being restricted by private firms, which seek to use their genomic knowledge for commercial development.

But the issues go beyond access. With companies racing to patent genes, any development of treatments or useful applications of genetic materials will risk being held hostage by exclusive ownership rights. At the very least, this is a disastrous cross-licensing nightmare in the making. As John Sulston, director of the Sanger Center, explains, "It will be an enormous ball and chain to medicine if, by the year 2003, control of every single gene is tied up by one company or another for twenty years. It seems obvious that, for the good of humanity, we should try to keep these things in the publicly exploitable domain."

We stand at a transformative moment. The issue of ownership of human genome information is highly polarized, and so far, advocates of open, unrestricted access are losing ground to private claims. Making the situation all the more brazen, the exclusive private claims are proliferating on the back of a huge amount of public investment in the project: by the time the map of the human genome is complete, U.S. taxpayers will have contributed nearly $3 billion toward the effort.

While genetic researchers race to build a preliminary map of the human genome, it is important to remember that this ongoing worldwide effort is little more than a prelude to the real challenge, namely, figuring out how genes work together to run the human body. This work will offer important benefits to humanity as well as a host of lucrative opportunities to develop specific drugs and other inventions.

In weighing public versus private ownership of the rudiments of the human genome map, we need to think carefully about the prospects for this resource in the future. Both scenarios offer viable alternatives. Neither will shut down the enterprise. And despite the often-repeated mantra championing incentives as a driver for technological development, it is important to remember that even without financial incentives, the project is of sufficiently compelling scientific interest that the genome will be fully mapped.

Meanwhile, though, the ownership arrangements we hammer out will profoundly influence the future. The prospect should not be compromised by today's quest for short-term private gain. After all, as Francis Collins, director of the National Center for Human Genome Research, explains it, making sense of today's human genome map "will be the topic of biomedical research for the next 100 years."

Casting the (Inter)Net

The Internet readily lends itself to hyperbole. Begun in 1969 as a project of the Pentagon's Defense Advanced Research Projects Agency (DARPA) to investigate computer networking technology, the Internet has exploded into an essential infrastructure for research, education, and entertainment. At a time when "giant information empires own everything else," the *New York Times Magazine* proclaimed in 1994, the Internet is "the most universal and indispensable network on the planet."

In fact, it is hard to say what is more exciting: the Internet's decentralized nature or its phenomenal growth, like a populist wildfire, to link millions of users throughout the world. So far, anyone with access to a computer, a modem, and a modest budget has been able to send and receive messages, as well as to read, copy, and distribute documents, manifestos, essays, and exposés. And the network is growing exponentially: at least 57 million Americans are connected to the Internet and on-line information services today, and the number of Web pages — now reported to be in the hundreds of millions — continues to expand.

But some of the most appealing aspects of the Internet's accessible smorgasbord of information are imperiled. The amount of high-quality, freely available on-line information has continued to grow, but already the best electronic data and publications are available only to registered users on a pay-per-use basis. In and of themselves, such metered-access arrangements pose few problems. But if a commercial imperative takes hold in cyberspace, it is likely to threaten formerly shared assets.

Powerful Internet players from MCI to Visa have set their sights on the creation of widespread on-line commerce. Most commentators argue that the success or failure of the Internet will hinge on the extent to which this commerce can successfully take root there. But in fact, the opposite may be more true. The utility of the Internet for our democracy will depend largely on the extent to which we can retain a healthy noncommercial domain within it so that information can continue to be shared, not just bought and sold.

In this sense, the juggernaut of commercial interest threatens to undermine and marginalize the decentralized and often fragile efforts to consider how the Internet can aid community, improve access to basic information services, and strengthen public participation in governance. Even the Clinton administration's highly touted plans for a National Information Infrastructure have been derailed by qualms over ownership rights. The debate is clearly illustrated by President Clinton's 1995 campaign pledge to hook up schools and libraries to the Internet as part of a "bridge to the twenty-first century." The president initially proposed that of all the nation's public schools, libraries, and rural health care facilities receive gratis a basic bundle of e-mail accounts and high-speed access to the World Wide Web, an inspired idea that would have taken advantage of some of the most promising aspects of the new technology. Under the plan, telephone companies, cable providers, and other telecommunications outfits would foot the bill.

But Clinton's plan was shouted down as unrealistic. *Business Week* in 1995 listed it as one of the ten worst ideas to come out of the presidential campaign between Clinton and Dole. As the magazine explained, the telecommunications firms "absolutely, positively hated the idea." Netscape counsel Peter

Harter wrote in response to the plan: "To fashion a subsidy system now, under the guise of extending the Internet to educational and medical institutions, would threaten government economic entanglement with the rapidly changing Internet on an unprecedented scale."

Despite the Clinton plan's unpopularity with industry, when it comes to the dissemination and development of knowledge, we have always had a subsidy system, and that is exactly what we need in the future. The Federal Communications Commission did work out a compromise in which telecommunications firms are mandated to establish discounted rates, 20 to 90 percent below their commercial rates, to schools and libraries. The arrangement, however, established guidelines only for connection charges. The bigger battle, over public access to content in cyberspace, is yet to come.

Some developments show the powerful potential for freely sharing information through this medium. As noted in Chapter 5, the National Library of Medicine decided in 1997 to make its huge MedLine database available on the Internet for free, in a move billed as a boon to the health of all Americans. And a small but ambitious program called Project Gutenberg seeks to place thousands of classic public-domain texts and reference materials on the Internet by the year 2001 to be freely disseminated.

But the Internet stands at a critical juncture: This new communication tool could benefit all citizens, or it could line the pockets of select groups, reinforcing existing disparities. Many of the Internet's biggest boosters are worried by what they see. "Without explicit, vigilant attention," says Michael Dertouzos, director of the Laboratory for Computer Science at MIT, for instance, "there is little question that the information infrastructure will increase the gap between rich and poor."

Sanctuaries, Zoning, and Antitrust

From soybeans to software, we have looked at the rising value of knowledge assets and the threat that exclusive private ownership of them poses to our social institutions. At the library, the university, and even the museum, more and more of our shared technological and cultural wealth is being auctioned off to private bidders. We have seen that disputes over the ownership of knowledge resources now fill court dockets and cause strains between nations. What can be done to preserve some categories of knowledge — raw data, essential techniques, and cultural treasures — so that they can be shared by all of us rather than controlled by a few for private gain?

For the most part, finite tangible goods lend themselves to free market arrangements. As we have seen, however, the vast conceptual realm does not naturally lend itself to competitive market-based trade. Rather, like the parts of our infrastructure — such as roads or clean drinking water — ideas and concepts must be open to all. The needs of the polity must take priority, or else the system of private ownership will — like Herman Daly's invisible foot — smash the public good to pieces. Our immediate challenge is to develop a vocabulary that can help frame our choices for protecting the shared conceptual commons. We need to develop a consensus supporting alternatives to the dysfunctional pseudo-market system that is now gaining ground. This may even require entirely new arrangements, such as several recent proposals to establish a wholly new regime of intellectual property protection for software. Most often though, it will simply require an enlightened triumph of democratic values over a metered market approach.

A good starting point is to look at the methods we have used successfully in the past to curb the excesses of unfettered development in more tangible realms. In particular, at this historic moment, it is worth reflecting on the saga of another generation's frontier.

In the second half of the 1800s, as the United States was still being settled, the sprawling nation epitomized the frontier spirit. The California gold rush drew fortune seekers to the West in record numbers as early as 1849, and the first transcontinental railroad was completed in 1869. But enormous tracts of land, especially in the country's interior, remained unclaimed in the public domain.

Apart from Native American populations, whose long stewardship of the nation's resources offers an important model, few American citizens in this period thought about the need to preserve land or natural resources. Most would likely have scoffed at the notion. When they looked at nature, they saw work to be done: wilderness to be conquered and settled, forests to be cut for timber, rivers to be tamed for power and irrigation.

It is not surprising, given the prevailing views of that frontier period, that the citizens of the youthful United States adopted an almost recklessly wasteful attitude toward the country's resources, leveling vast regions of forests and decimating the habitats of many species of wildlife, among many other offenses to the natural well-being of the land, water, and air.

Slowly but surely, though, a small countervailing current arose in reaction to the excesses and waste of this period of unfettered development. Persuaded by the writings of Emerson and Thoreau, for instance, a small but significant number of people began to appreciate the value of wilderness as something other than a resource to be exploited.

Some of these ideas found a practical foothold in the summer of 1870, when a group of explorers led by General Henry Washburn mounted an expedition in the largely unexplored Yellowstone region, part of what was then known as the Montana territories. According to the journal of expedition member Nathaniel Pitt Langford, the group was enthralled by the area's geothermal hot springs, geysers, waterfalls, and canyons. How the plan for a national park was spawned is a matter of continued dispute among historians. But there is little question that the enthusiasm for this vast tract that the Washburn expedition members expressed upon their return helped sway President Ulysses S. Grant to sign the Yellowstone Park Act in 1872, establishing the nation's first national park. The idea of preserving wilderness for the general public — not to mention some 2 million acres in the heart of the nation — was radical and unprecedented.

It was not an easy sell in Congress and, for the first several decades of its existence at least, Yellowstone National Park was often dismissed as a failure. But the seed had been sown, and eventually the idea of public preserves would reap a rich harvest of gratitude from future generations.

The example of Yellowstone National Park inspired a man who would become one of the country's best-known naturalists, John Muir, founder of the Sierra Club and often known as the father of the country's national parks. Notably, Muir won the ear of President Teddy Roosevelt around the turn of the century. After Roosevelt accompanied Muir on a camping trip to a grove of giant sequoias in California in 1903, he waxed as dreamy and romantic about the importance of the wilderness as Muir often sounded, writing: "Lying out at night under those giant sequoias was like lying in a temple built by no hand of man, a temple grander than any human architect could build, and I hope for the preservation of the groves of giant trees simply because it would be a shame to our civilization to see them disappear."

Roosevelt's enthusiasm for protecting wilderness for posterity grew significantly over the course of his administration. Ultimately, Roosevelt tripled the amount of land in national forests, withheld 65 million acres from the coal companies, doubled the number of national parks, and set up fifty-three animal refuges and seventeen national monuments, including the Grand Canyon.

The idea of preserving unique natural settings took off in the first decades of the 1900s. Around this period, for instance, Frederick Law Olmsted increased public interest in the creation of urban parks, which he designed throughout the Northeast, including Central Park in New York City and the so-called Emerald Necklace of conjoined parkland that winds through Boston. George W. Field's appreciation for birds would lead him to found the Massachusetts Audubon Society by turning over to it his 225-acre estate in Sharon, Massachusetts.

Today the U.S. National Park Service presides over some 80 million acres that are held in public trust. Yellowstone National Park, once an idealistic vision, now plays host to roughly 3 million visitors annually. Equally noteworthy, some 140 nations have emulated the model started at Yellowstone.

There is much to learn from these episodes. Most would agree that we are vastly better off for these various conservation efforts, which represented a bold and unexpected departure from the spirit of development that prevailed at the height of the era of industrialization. Of course, natural resources are distinct from the intangible assets now confronting us. But the lessons are surprisingly similar. When unfettered ownership of the frontier threatened to erode some of its most cherished features, public action was needed as a corrective.

The question is, do we have the farsighted vision to put ourselves in the shoes of the Washburn expedition members as we survey the terrain of knowledge assets? Now, as then, the stakes are high. Today we must begin to define the public domain in the realm of intellectual property, the conceptual commons that will prove indispensable to our future progress and growth.

SANCTUARIES

Most people would probably be aghast if a developer like Donald Trump proposed to buy the Grand Canyon in order to build a new amusement park. But what if he proposed to buy the Grand Canyon for a handsome price and keep it in its natural state? Would it make a difference if this natural wonder were in private hands? What if the trade-off included a hefty admission price or, as in the current debate over some public lands in Alaska, the right to mine its resources?

For most of us, the public trust status of the Grand Canyon and other national sanctuaries has come to be seen as sacrosanct. Is it possible that certain knowledge assets need to be similarly protected, designated as off-limits for individual, private ownership? If so, the map of the human genome would be an excellent starting point. For a variety of reasons — ranging from the pragmatic to the morally based — the human genome appears as a stunningly powerful resource deserving of a special status. But how do we know a conceptual Grand Canyon when we see one? What should be the criteria for establishing such preserves in the conceptual realm? One useful principle is the need to protect our infostructure — those seminal ideas, standards, languages, and tools needed by all in a given field to do their jobs.

In the realm of tangible property, we have often allowed companies monopoly status to build infrastructures, such as telephone or electric lines to people's homes. The infostructure, however, is not made up of tangible items like roads but rather of intangible ones, more akin to rules of the road. Players in the knowledge-based economy are coming to realize that, whenever possible, the infostructure must be freely shared. Standards are essential to the development of new technologies, whether we are talking about telecommunications or even typewriter keyboards. To be successful, our machines must be able to communicate flawlessly with one another. Whenever possible, we need to establish and build on collectively agreed-upon formats, from the designated way pilots speak to air traffic controllers to the way appliances plug into wall sockets.

While at times tempting for the speedy development it promises, it is ultimately unproductive to allow individuals to leverage this type of knowledge. By this token, for instance, the recognition tones used by fax machines to connect with one another should have been seen as a piece of the infostructure, and Jerome Lemelson — regardless of the legitimacy of his contribution — should have been denied the right to exact royalty payments from every fax machine manufacturer.

The early history of the national park system highlights the conflicting pressures we are likely to face in coming years. In the early 1900s, the decision to preserve vast acreage often directly conflicted with short-term development goals. In one famous fight, the industrialist and millionaire Andrew Carnegie derided John Muir's stubborn opposition to a plan to build a dam in the Hetch Hetchy Valley within Yosemite National Park. As Carnegie put it, "It is too foolish to say that the imperative needs of a city to a full and pure water supply should be thwarted for the sake of a few trees or for scenery, no matter how beautiful it might be." Carnegie would eventually win the battle to dam the Hetch Hetchy valley, but ultimately the imperative of halting the encroachment of private development on these lands held sway at national parks around the country. As J. Horace McFarland, president of the then-powerful American Civic Association, argued at the formal establishment of the National Park Service in 1916, "These parks did not just happen. They came about because earnest men and women became violently excited at the possibility of these great assets passing from public control."

ZONING

If sanctuaries offer a way of securing certain important categories of knowledge for open, public use, other forms of knowledge may well need regulation akin to zoning, which allows private ownership but restricts certain specific uses so as not to interfere with the rights of others or erode agreed-upon desirable features of a given area or neighborhood.

Vivian Weil and John Snapper, two philosophers at Illinois Institute of Technology, have pointed out that our conception of ownership is less than straightforward. To be sure, the owner of an item has some say over who may use it and how. There is, however, no single form of control that holds in all ownership claims. As Weil and Snapper note, a homeowner is said to own a house; he or she can live in it, rent it, or sell it, but he or she cannot necessarily build a tower on top of it or, in some historic neighborhoods, even choose what color to paint it. Similarly, a book owner can lend, sell, even burn the book. But he or she is not permitted to copy any but the smallest portion of it without express permission of the author.

Zoning is used as a tool to protect key aspects of a community from undesired acts of individual owners. Zoning laws codify our collective understanding that the presence of even one industrial plant can irrevocably alter the character of a residential neighborhood, that one modern high-rise in a historic district can similarly erode its qualities. In a similar fashion, as we seek to create more equitable arrangements in the knowledge-based economy, a kind of zoning can be used to mandate some degree of access, reduced fees, or other guarantees, even in those cases where we allow the private ownership of intellectual property. In the tangible realm, arrangements such as conservation easements, protection of a view, or provision of a public right of way to a beach are relatively common.

A zoning-like approach could effectively be used to curb the ills of exclusivity — perhaps the most overused and undesirable feature of ownership over knowledge assets. In the conceptual realm, exclusivity is like a disease that infects everything it touches. And yet we are seeing more and more exclusive ownership arrangements as savvy players work in myriad ways to secure a stream of profit from the knowledge assets they control.

A kind of conceptual zoning arrangement, for example, could be used to mandate compulsory licensing of certain kinds of patents in fields such as health care, where public interest concerns are strong. Similar kinds of arrangements could require exemptions to exclusive ownership rights in cases where the desired knowledge is sought for noncommercial purposes.

Perhaps some form of zoning can be adapted particularly effectively to the realm of cyberspace, establishing rules of access and fee structures that apply differently to some communities of users, such as schools, public institutions, or low-income communities. In all such efforts, the key is to begin a high-profile dialogue that establishes and legitimizes the notion of collective rules to govern the actions of private owners within certain defined realms.

ANTITRUST

As we have seen in many disparate fields, private ownership claims can place an unacceptable amount of power in the hands of knowledge moguls and technological titleholders. As we think about safeguarding a productive system, an important principle should be to safeguard fair competition. Given the excesses of the existing patent system, we need to seriously reconsider — at the very least — the ease with which the government hands out sanctioned monopolies on conceptual terrain. In the realm of intellectual property, as elsewhere, monopolies should be a refuge of last resort, not a dominant piece of our industrial policy, as they are in the patent system.

As we have seen, monopolies over ideas do not necessarily improve productivity and, in fact, can stifle development and innovation. In addition, they raise profound questions of equity and legitimacy, as illustrated by the U.S. Patent Office's record in granting exclusive rights to overly broad swaths of conceptual terrain and in not distinguishing clearly between discoveries of natural phenomena and actual inventions.

Along these lines, several close watchers have begun to focus on the size of the conceptual monopoly chunks issued by the Patent Office. It is becoming clear that some of the worst problems occur at both ends of the range. When overly broad conceptual patents are awarded, competitive development in a field is blocked. Problems also arise when patents are issued for chunks that are too small, causing a stifling, layer-cake effect in which a morass of licenses and royalty payments must be negotiated before anyone can accomplish anything.

Our government has historically tried to intervene in situations where the threat of monopoly ownership has lent an unfair advantage in the marketplace. From utilities to phone service, technological infrastructure has long been characterized by monopoly arrangements and by efforts to battle their excesses. The problem today — as the unfolding Microsoft antitrust case illustrates — is that we have yet to establish a clear sense of what antitrust means in the knowledge economy. We need to revitalize our notion of antitrust law to explicitly restrict monopolies over the infostructure — monopolies that represent some of the most potentially dangerous concentrations of power we have seen yet.

When the market fails, as it often does in the conceptual realm, the best model we have to rely upon is called democracy. Democratic processes and institutions can tame the excesses of monopoly ownership, insisting upon arrangements that feature pooled risks and shared benefits. This is the system that brought us our unique public institutions in the first place — that forged the idea of a library open to all and championed the notion of public education and equal opportunity. Even the U.S. Postal Service, though often the butt of jokes, can be seen to incorporate our democratic principles. The postal service is self-supporting, but it uses cross-subsidies to retain a low-cost, first-class domestic stamp that allows participants to send a letter near or far for a set price.

What is important about systems like these is their mandate to serve a generous conception of the public. Such arrangements result in an aggregate kind of fairness that absorbs the additional cost of serving a rural postal route or offering a less-trafficked bus route. We need to draw upon this generous, democratic conception of the public as we establish new norms for apportioning rights to knowledge assets.

At this key juncture, the task before us all, as participants in a democratic system, is nothing less than to recapture and redefine our rights to our shared cultural inheritance, setting forth democratic principles to guide the use of the conceptual commons and guard against its private capture. Knowledge is intimately linked with culture, education, and governance, as the nations of the world have recognized ever since they endorsed the Universal Declaration of Human Rights at the United Nations in 1948. That statement reads, in part, "Everyone has the right to freedom of opinion and expression; this right includes freedom to hold opinions without interference and to seek, receive and impart information and ideas through any media and regardless of frontiers." In the postwar climate, the right to "receive and impart information and ideas" was undoubtedly conceived of as an antidote to censorship in totalitarian regimes. But it serves as a powerful ideal for unimpeded access to knowledge in an era in which that access is increasingly imperiled.

Ultimately, the principles we establish to assure public access to the conceptual commons matter more than the speed of technological development or the enormous sums of money that can be leveraged from the private control of knowledge assets. They matter because, whether we are studying the human genome, surfing the Internet, or displaying a digital image of the Mona Lisa, we in a democratic society have the right to shape our own future. And, whether we face a threat to our public lands, to the availability of clean water, or to access to information, we have the right and the responsibility to protect the precious assets in the public domain. If we fail, we will relinquish, erode, and eventually lose our public voice. And in a democratic society, even a wealthy one like ours, this may be our most precious asset of all.

Notes
Acknowledgments
Index

Notes

PREFACE

PAGE

xi *the nation's first public library.* John A. Peters and Nina C. Santoro, *A History of America's First Public Library at Franklin, Mass., 1790–1990* (Franklin, Mass.: Franklin, Mass., Bicentennial Committee, 1990).

xii *volume of town history.* Mortimer Blake, *A History of the Town of Franklin, Massachusetts* (Franklin, Mass.: Committee of the Town, 1878).

"sense is preferable to sound." James C. Johnston, Jr., *Images of America: Franklin* (Dover, N.H.: Arcadia Publishing, 1996).

xiii *Harvard College's book collection.* Information from Harvard University press office, July 1998.

Franklin had started a membership library. Abigail A. Van Slyck, *Free to All: Carnegie Libraries and American Culture* (Chicago: University of Chicago Press, 1995).

xiii *16,000 local branches.* American Library Association, "The 1997 National Survey of U.S. Public Libraries," data sheet, 1997.

Horace Mann. Peters and Santoro, *History,* p. 12.

1. GOLD RUSH IN THE IDEA ECONOMY

3 *"nineteenth-century robber barons."* James Boyle, "Sold Out," *New York Times,* Mar. 31, 1996, p. 15.

portion of the human genome. Interview with Robert Cooke-Deegan, staff scientist, National Academy of Sciences, Mar. 1998.

4 *Shakespearean scholars have even gone to court.* David Remnick, "Hamlet in Hollywood," *New Yorker,* Nov. 20, 1995, pp. 66–83.

"The most important form of property." Alvin and Heidi Toffler, *Creating a New Civilization* (Atlanta: Turner Publishing, 1994), p. 67.

blue-collar manufacturing workers. Alvin Toffler, *Powershift: Knowledge, Wealth, and Violence at the Edge of the Twenty-first Century* (New York: Bantam Books, 1990), p. 71.

"New Economy of Ideas." Allan Sloan, "The New Rich," *Newsweek,* Aug. 4, 1997, p. 51.

5 *Texas Instruments has earned a larger portion.* Norm Alster, "New Profits from Patents," *Fortune*, Apr. 25, 1988, pp. 185–90.

6 *required patent applicants to submit physical models.* Cathleen Schurr, "Two Hundred Years of Patents and Copyrights," *American History Illustrated* (July–Aug. 1990), p. 63.

ten thousand visitors per month. Ibid., p. 64.

"actionable knowledge." Esther Dyson, "Intellectual Value," *Release 1.0*, Dec. 1994, EdVenture Holdings.

7 *a particular improved mousetrap.* Testimony by Wallace Judd, U.S. Patent and Trademark Office, "Public Hearing on Use of the Patent System to Protect Software-Related Inventions," San Jose, Calif., Jan. 26–27, 1994.

"the crucial right to exclude others." Interview with Charles Van Horn, Mar. 1995.

our patent system has not adapted. See, for instance, Pamela Samuelson, "Innovation and Competition: Conflicts over Intellectual Property Rights in New Technologies," in Vivian Weil and John W. Snapper, eds., *Owning Scientific and Technical Information: Value and Ethical Issues* (New Brunswick, N.J.: Rutgers University Press, 1989), pp. 169–92; John Perry Barlow, "Selling Wine without Bottles: The Economy Mind of the Global Net," in Peter Ludlow, ed., *High Noon on the Electronic Frontier* (Cambridge, Mass.: MIT Press, 1996); and James Boyle, *Shamans, Software, and Spleens: Law and the Construction of the Information Society* (Cambridge, Mass.: Harvard University Press, 1996).

"the most unreal estate." John Perry Barlow, "The Economy of Ideas: A Framework for Rethinking Patents and Copyrights in the Digital Age," *WIRED Online* (1993), p. 4.

8 *the courts are conflicted.* See, for example, Anne Branscomb, *Who Owns Information: From Privacy to Public Access* (New York: Basic Books, 1994).

inserting particular genes into cotton. Seth Shulman, "A New King Cotton?" *Technology Review*, July 1994, pp. 16–17.

ex vivo human gene therapy. See, for example, Jeff Lyon and Peter Gorner, *Altered Fates: Gene Therapy and the Retooling of Human Life* (New York: W. W. Norton, 1995); "FTC Accord in Ciba Geigy/Sandoz Merger," Federal Trade Commission press release, Dec. 17, 1996.

controversial experiment in 1995. Raul J. Cano and M. K. Borucki, "Revival and Identification of Bacterial Spores in 25- to 40-Million-Year-Old Dominican Amber," *Science*, May 19, 1995, pp. 1060–64. For dispute of the findings, see Andrew T. Beckenbach and F. G. Priest, *Science*, Dec. 22, 1995, p. 2015.

9 *monopoly ownership Cano won is "extremely broad."* Teresa Riordan, "A Molecular Biologist Receives Protection for Discoveries That Could Be Millions of Years Old," *New York Times*, March 31, 1997, p. D2.

"the patent system has become a lottery." Editorial, *Forbes*, Mar. 29, 1993.

the case of Ashleigh Brilliant. David D. Kirkpatrick, "Brilliant Minds May Think Alike, but Brilliant Lines Can Cost You," *Wall Street Journal*, Jan. 27, 1997, p. B1.

Chicago-based Qualitex Company. Joan Biskupic, "Product Colors Receive Trademark Protection," *Washington Post*, Mar. 29, 1995, p. A9.

10 *Justice Stephen G. Breyer asked.* Ibid.

a hobby magazine for Barbie enthusiasts. Phillip M. Perry, "Fair Use or Infringement?" *Folio: The Magazine for Magazine Management,* Nov. 1, 1997, p. 101.

Sir Roger Penrose. "Artifact: Sir Roger Penrose Filed a Patent Infringement Lawsuit," *Time,* May 5, 1997, p. 26.

athletes may have the legal right. As reported in "Yo! He Owns That Move," *Sports Illustrated,* May 27, 1996, p. 16. The article cites *National Law Journal,* May 20, 1996.

a Texas court ordered Evan Brown. "Who Owns What's Inside Your Head?" *Information Week,* July 14, 1997. See also John S. Pratt and Peter Dosik, "Whose Idea Is It?" *National Law Journal,* Oct. 20, 1997, p. C16.

11 *"If a janitor."* George Bunt quoted in Geanne Rosenberg, "An Idea Not Yet Born, but a Custody Fight," *New York Times,* Sept. 8, 1997, p. D3.

"the public domain will be so diminished." Boyle, "Sold Out," p. 15.

a patent on Kirchoff's Law. As recounted in testimony by Richard Stallman, U.S. Patent and Trademark Office, "Public Hearing on Use of the Patent System to Protect Software-Related Inventions," San Jose, Calif., Jan. 26–27, 1994.

a patent claim over two large prime numbers. See U.S. Patent No. 5,297,206, "Secure File Transfer System and Method," issued to Roger Schlafly and Michael J. Markowitz, Mar. 22, 1994.

12 *called it "outrageous."* Pamela Samuelson quoted in Simson Garfinkel, "A Prime Argument in Patent Debate," *Boston Globe,* Apr. 6, 1995.

"crisis in intellectual property law." Samuelson, "Innovation and Competition," in Weil and Snapper, *Owning Scientific and Technical Information,* pp. 169–92.

2. THE NEW WEALTH OF NATIONS

13 *first patent on an altered life form.* See *Diamond v. Chakrabarty,* 447 U.S. 305 110 S. Ct. 2204, and 447 U.S. 303, 308–9 (1980). Diamond was Sidney A. Diamond, then commissioner of the U.S. Patent and Trademark Office, whose patent examiners had initially rejected Chakrabarty's patent application.

IBM was granted 1,724 U.S. patents. Press release, "IBM Leads in U.S. Patents for Fifth Consecutive Year, Capping 1997's Technological Breakthroughs," *Business Wire,* Jan. 12, 1998. For annual listing, see Intellectual Property Owners, "Top 200 Organizations Receiving U.S. Patents in 1996," statistical sheet, 1997.

close to $1 billion. Mo Krochmal, "IBM Collects on Intellectual Property," *TechWeb News,* Nov. 6, 1997.

14 *Internet software firm Netscape Communications.* See Elizabeth Corcoran, "Netscape Comes Down to Earth," *Washington Post Weekly Edition,* Jan. 19, 1998, p. 19. Heather Green, "Has Netscape Hit the 'Innovation Ceiling'?" *Business Week,* Jan. 19, 1998.

"Instantaires." James Collins, "Instantaires," *Time,* Feb. 19, 1996, p. 43.

14 *"from the business of making things."* Allan Sloan, "The New Rich," *Newsweek,* Aug. 4, 1997, p. 51.

"we are at a major moment." William Wresch, *Disconnected: Haves and Have-Nots in the Information Age* (New Brunswick, N.J.: Rutgers University Press, 1997), p. 96.

Adam Smith's hallowed eighteenth-century notion. Adam Smith, *The Wealth of Nations* (1776; reprint: Chicago: University of Chicago Press, 1976). See also "The Point of Patents," *The Economist,* Sept. 15, 1990, p. 19.

Distinguished Professor of Knowledge. James Sterngold, "Professor of Knowledge Is Not an Oxymoron," *New York Times,* June 1, 1997, p. D5.

"I liked it better when we made steel." Ibid.

15 *We build houses and offices and factories from information."* Walter Wriston, *The Twilight of Sovereignty* (New York: Charles Scribner's Sons, 1992), p. 23.

"sit upon a 'knowledge base.'" Alvin and Heidi Toffler, *Creating a New Civilization* (Atlanta: Turner Publishing, 1994), p. 35.

Brookings Institution surveyed. As reported in H. Garrett DeYoung, "Thieves among Us: If Knowledge Is Your Most Important Asset, Why Is It So Easily Stolen?" *Industry Week,* June 17, 1996, p. 12.

16 *"the great end of life."* Thomas Henry Huxley, *Science and Christian Tradition* (New York: D. Appleton and Co., 1898).

"the most influential woman." Claudia Dreifus, "The Cyber-Maxims of Esther Dyson," *New York Times Magazine,* July 7, 1996, p. 16.

"distribute intellectual property free." Esther Dyson, "Intellectual Value," *Release 1.0,* Dec. 1994, EdVenture Holdings.

those who attend $500-a-plate dinners. Ibid.

17 *"the virtual value chain."* Jeffrey F. Rayport and John J. Sviokla, "Exploiting the Virtual Value Chain," *Harvard Business Review,* Nov.–Dec. 1995, p. 83.

o*which associates to make partner.* Interview with Jim Rebitzer, Mar. 1997. See also Jim Rebitzer and Lowell Taylor, "Efficiency Wages and Employment Rents: The Employer Size Wage Effect in the Job Market for Lawyers," unpublished manuscript.

too little information will be generated. Arrow is discussed in James Boyle, Shamans, Software, and Spleens: Law and the Construction of the Information Society (Cambridge, Mass.: Harvard University Press, 1996), pp. 41–42. See Kenneth Arrow, "Economic Welfare and the Allocation of Resources for Invention," in *Rate and Direction of Inventive Activity: Economic and Social Factors* (Washington, D.C.: National Bureau of Economic Research, 1962), pp. 609, 617.

weak intellectual property regime. Boyle, *Shamans,* p. 42. See Eugene F. Fama and Arthur B. Laffer, "Information and Capital Markets," *Journal of Business* 44 (1971): 289–98.

"this new world will distribute." Dyson, "Intellectual Value," p. 8.

18 *"the attempt to block such changes."* Toffler and Toffler, *Creating a New Civilization,* p. 106.

75 percent of the world's sales. See testimony of William Neukom, vice president of law and corporate affairs, Microsoft Corp., in U.S. Patent and Trademark Office, "Public Hearing on Use of the Patent System to Protect Software-Related Inventions," San Jose, Calif., Jan. 26–27, 1994, transcript, p. 66.

18 *losing tens of billions of dollars.* See Benjamin Kang Lim, "China, U.S. Sign Anti-Piracy Pact, Avert Trade War," Reuters, Feb. 26, 1995, citing U.S. government officials' complaints of China's piracy of U.S. intellectual property amounting to nearly $1 billion annually.

"the patent is a sovereign device." Interview with Lawrence Goffney, Mar. 1995. See also Seth Shulman, "Patent Medicine," *Technology Review,* Nov.–Dec. 1995, pp. 29–36.

19 *the ownership of ever more intangible assets.* Richard Barnet and John Cavanagh, *Global Dreams* (New York: Simon and Schuster, 1994), pp. 335–36. For discussion of patents in developing countries, see p. 354.

Thailand has tried to protect its native healers. "Thailand: Oxford Dons or Plant Predators?" *RAFI Communiqué,* Sept.–Oct. 1995, p. 3.

France has also taken an outspoken stance. See the comments of Prime Minister Lionel Jospin, in William Drozdiak, "Down with Yankee Dominance: America's Post–Cold War Influence on the World Leaves Its Allies and Neighbors Cold," *Washington Post Weekly Edition,* Nov. 24, 1997, p. 15.

to occupy the parliament in 1996. "Biopiracy and Indigenous Knowledge," *Yearend Roundup,* Rural Advancement Fund International (RAFI), Dec. 1996.

500,000 farmers in southern India demonstrated. See Hope Shand, "Patenting the Planet," *Multinational Monitor,* June 1994, p. 9. See also Chirag Mehta, "The Seed Satyagraha: Indian Farmers and Global Capital Face Off," *Dollars and Sense,* Sept.–Oct. 1994, p. 25.

a 1997 letter to U.S. Secretary of State Madeleine Albright. The open letter to Albright was organized in part by the Institute for Agriculture and Trade Policy, Minneapolis, June 20, 1997.

20 *American officials willfully ignored British patents.* See Lester C. Thurow, "Patents and Pirates," *Boston Globe,* June 18, 1996, p. 42.

"the largest sanctions." Kantor's remarks in a CNN interview are quoted in Donna Smith, "U.S. Prepares Trade Sanctions against China," Reuters, Feb. 4, 1995. See also David E. Sanger, "This One Just Might Be a Real Trade War," *New York Times,* May 19, 1996, p. 6.

the Clinton administration threatened trade sanctions. Martin Crutsinger, "U.S. Warns Nations on Copyright Law," Associated Press, Apr. 30, 1997.

21 *what some have termed a liminal space.* Peter Lyman makes this point about libraries in particular. See Peter Lyman, "What Is a Digital Library? Technology, Intellectual Property, and the Public Interest," *Daedalus* (special issue, "Books, Bricks, and Bytes"), Fall 1996, p. 3.

3. STAKING OUR CLAIM

22 *the need of some shark species.* John Perry Barlow, "Selling Wine without Bottles: The Economy Mind of the Global Net," in Peter Ludlow, ed., *High Noon on the Electronic Frontier* (Cambridge, Mass.: MIT Press, 1996), p. 19.

23 *"nonrival consumption."* See Patrick Croskery, "The Intellectual Property Literature: A Structured Approach," in Vivian Weil and John W. Snapper, eds., *Owning Scientific and Technical Information: Value and Ethical Issues* (New Brunswick, N.J.: Rutgers University Press), p. 270. See also Paul Romer, "Two Strategies for Economic Development: Using Ideas and Producing Ideas," working paper no. 4, Canadian Institute for Advanced Research, Program in Economic Growth and Policy, Toronto, May 1992, pp. 18–19, as discussed in Robert Kuttner, *Everything for Sale* (New York: Knopf, 1997), pp. 199–200.
"If more people use a software program." Interview, "Richard Stallman Discusses His Public-Domain UNIX-Compatible Software System with Byte Editors," *Byte*, July 1986.

24 *"The faster knowledge is spread."* Lester C. Thurow, "Patents and Pirates," *Boston Globe*, June 18, 1996, p. 42.
"the value of a piece of scientific work." Norbert Wiener, *Invention: The Care and Feeding of Ideas* (1954; posthumously published: Cambridge, Mass.: MIT Press, 1993), p. 153.

25 *plagiarism or piracy.* Pamela Samuelson makes a similar point in "Innovation and Competition," in Weil and Snapper, *Owning Scientific and Technical Information*, pp. 170–71.
the free-rider problem. See, for example, Arthur Kuflik, "Moral Foundations of Intellectual Property Rights," in Weil and Snapper, *Owning Scientific and Technical Information*, p. 231.

26 *"a Gibraltar of technological inheritance."* Gar Alperovitz, "Distributing Our Technological Inheritance," *Technology Review*, Oct. 1994, p. 31.
taxpayers contribute some $60 billion annually. See "The Leverage of Federal Research," editorial, *New York Times*, May 18, 1997. See also Don E. Kash and Robert W. Rycroft, "Nurturing Winners with Federal R&D," *Technology Review*, Nov.–Dec. 1993, pp. 59–64.

27 *New York City has hired International Management Group.* Harry Hurt III, "Parks Brought to You by . . . Can IMG Do for Cities What It Has Done for Tiger Woods and the Pope?" *US News & World Report*, Aug. 11, 1997, p. 45.
Coca-Cola and PepsiCo have paid schools. Ibid. See also Elliott D. Sclar, "Public-Service Privatization: Ideology or Economics?" *Dissent*, Summer 1994, p. 329; and Mike West, "Coca-Cola High," *Progressive*, Nov. 1997, p. 26; Constance L. Hays, "Be True to Your Cola, Rah! Rah!" *New York Times*, Mar. 10, 1998.
"the tragedy of the commons." Garrett Hardin, "The Tragedy of the Commons," *Science* 162 (1968): 1243–48.
"invisible foot." See Herman Daly, "Towards a Stationary-State Economy," in John Harte and Robert Socolow, eds., *Patient Earth* (New York: Holt, Rinehart and Winston, 1971).

28 *"The press and the public are being slowly blinded."* Bill Kovach, "When Public Business Goes Private," op-ed article, *New York Times*, Dec. 4, 1996.
"The Capitalist Threat." Atlantic Monthly, Feb. 1997, p. 48.

29 *"Does anyone believe that Central Park would exist."* Fallows quoted in Daniel Schorr, "Public Funds for Public Culture," op-ed article, *Boston Globe,* Mar. 10, 1997.

no "true man of science." As quoted in Schurr, "Two Hundred Years of Patents and Copyrights," *American History Illustrated,* July–Aug. 1990, p. 63.

"a popular government, without popular information." Letter from James Madison to W. T. Barry, Aug. 4, 1822, reprinted in Saul K. Padover, ed., *The Complete Madison* (New York: Harper, 1953), p. 337.

4. THE NEW MEDICAL LICENSES

33 *a surgeon in Arizona had recently patented.* U.S. Patent No. 5,080,111, "Method of Making Self-Sealing Episcleral Incision," issued to Samuel L. Pallin, Jan. 14, 1992.

34 *"I had nearly slammed my fist."* Interview with Jack Singer, July 1997.

half a million dollars in legal fees. Interview with Jack Singer, Sept. 1997.

concern from sixteen separate organizations. As listed in "President Signs Medical Patent Bill; Physicians Freed from Threat of New Medical Procedure Patent Lawsuits," press release, American Society of Cataract and Refractive Surgery, *PR Newswire,* Oct. 1, 1996.

35 *"If the Heimlich maneuver had been patented."* Noonan quoted in Robert L. Lowes, "Are You Stealing from Other Doctors?" *Medical Economics,* Mar. 11, 1996, p. 198.

"the thousands of different surgical procedures." Robert Portman, "Patenting Medical and Surgical Procedures Is Threatening Medical Progress," *Policy Options,* May 1996, p. 31.

"It's part of the Hippocratic Oath." Robert McAfee, as quoted in Sally Squires, "AMA Condemns Patents for Medical Procedures," *Washington Post,* June 20, 1995.

36 *One hundred per month.* See Lowes, "Are You Stealing from Other Doctors?"

"Proliferating medical-procedure patents." Ibid., p. 195.

owns the rights to a basic technique for suturing. U.S. Patent No. 4,328,805, "Method of Suturing the Organs of the Gastrointestinal Tract," issued to Ernest M. Akopov and Petr M. Postolov, May 11, 1982; U.S. Patent No. 3,640,279, "Skin Graft Cutting Method and Machine," issued to Warren F. Brown, Feb. 8, 1972; U.S. Patent No. 5,546,964, "Method for Treating a Bleeding Nose," issued to Sven-Eric Stangerup, Aug. 20, 1996; U.S. Patent No. 5,709,869, "Method for Treating Nerve Injury Pain Associated with Shingles," issued to Harry Hind, Jan. 20, 1998.

37 *technique for determining the sex of a fetus.* U.S. Patent No. 4,986,274, "Fetal Anatomic Sex Assignment by Ultrasonography during Early Pregnancy," issued to John D. Stephens, Jan. 22, 1991.

"It's like saying you have a secret method." Interview with Chris Merritt, Dec. 1997.

37 *"naive and out of date."* Stephens quoted in Lowes, "Are You Stealing from Other Doctors?" p. 213.

"Everybody can start claiming ownership." Interview with Jack Singer, July 1997.

38 *only after Singer wrote an article.* See Jack Singer, "Frown Incision for Minimizing Induced Astigmatism after Small Incision Cataract Surgery with Rigid Optic Intraocular Lens Implantation," *Journal of Cataract and Refractive Surgery,* July 1991, pp. 677–88.

As Pallin tells the story. See Jodie Snyder, "A Patent for Eye Surgery?" *Phoenix Gazette,* Apr. 4, 1995, p. A1.

"yesterday's news." Ibid., p. A1.

patents were entirely and explicitly prohibited. Lowes, "Are You Stealing from Other Doctors?" p. 196.

39 *"What could a doctor do?"* Annas quoted in Edward Felsenthal, "Medical Patents Trigger Debate among Doctors," *Wall Street Journal,* Aug. 11, 1994, p. B1.

"it is not surprising." Ibid.

"medicine is a capitalist endeavor." "Ethics Debate Springing Up over Patenting Procedures," CNN *Early Edition,* Andrea Arceneaux interviews Dr. Jack Singer and Dr. Samuel Pallin, June 20, 1995.

40 *Pallin brought formal legal action.* Felsenthal, "Medical Patents Trigger Debate," p. B1.

enough to secure a legal victory. Mark Bloom, "Patent Claim on Sutureless Cataract Incision: No Way," *Physicians Weekly,* May 20, 1996.

"Medical advances need to be shared." "Dear Colleague" letter, July 18, 1996, supplied by Rep. Greg Ganske's office. See also Greg Ganske, "Medical Procedure Patents Put Patients at Risk," *Roll Call,* Sept. 16, 1996, p. 15.

41 *A 1995 AMA position paper.* "Patenting of Medical Procedures (Informational Report)," Report 1 of the American Medical Association Council on Ethical and Judicial Affairs (A-95), John Glasson, M.D., chair, 1995.

"struck terror into the heart." Jeffrey I. D. Lewis, "No Protection for Medical Processes," *New York Law Journal,* Mar. 10, 1997.

"one of the country's greatest assets." Ibid.

42 *The final legislation.* "A Bill to Limit the Issuance of Patents on Medical Procedures," H.R. 1127, 104th Cong., 1st sess., Mar. 3, 1995. Final bill amends United States Code, sec. 287, title 35.

43 *a bill expanding patent rights in biomedicine.* Clinton signed S. 111 on Nov. 1, 1995. See "Statement by President Clinton on Signing of Biotechnology Process Patents Protection Bill," *U.S. Newswire,* Nov. 1, 1995. For more on that bill, see also "Biotechnology Process Patents," *Congressional Record,* Oct. 17, 1995, p. S15221.

44 *rare type of genetic disorder.* Technical details are drawn largely from "Results from First Human Gene Therapy Clinical Trial," press release, National Institutes of Health, Oct. 19, 1995.

"We're absolutely thrilled." Blaese quoted in Dolores Kong, "Study: First Gene Therapy a Success," *Boston Globe,* Oct. 20, 1995.

45 *the filing by Anderson.* See Susan Jenks, "NIH Wins Broad Patent on Human Gene Therapy," *Journal of the National Cancer Institute* 87 (1995): 569.

"Deep disbelief." Glorioso quoted in Helen Gavaghan, "NIH Wins Patent on Basic Technique Covering All *Ex Vivo* Gene Therapy," *Nature* 374 (Mar. 30, 1995): 393.

"What we did was to provide proof in principle." Anderson quoted ibid.

"beyond comprehension." This and other quotes by Miller are from my interview with Dusty Miller, July 1997.

46 *authoritative book.* Jeff Lyon and Peter Gorner, *Altered Fates: Gene Therapy and the Retooling of Human Life* (New York: W. W. Norton, 1995), p. 160.

"I wound up in shock." Culver quoted ibid., pp. 246–47.

47 *editorial board of the new field's leading journal.* The journal is *Human Gene Therapy,* Mary Ann Liebert, Inc., publishers.

"I think the public's been ripped off here." Miller quoted in Tom Paulson, "Who Owns the Rights to Gene Research?" *Seattle Post-Intelligencer,* Feb. 14, 1997.

48 *"foster collaborations."* Barrett quoted in Kathryn Brown, "Major Pharmaceutical Companies Infuse Needed Capital into Gene Therapy Research, *The Scientist* 9, no. 22 (Nov. 13, 1995): 10.

Sandoz bought GTI. Lawrence M. Fisher, "Sandoz Buying Genetic Therapy for $295 Million," *New York Times,* July 11, 1995, p. D4.

$2.5 million in venture capital. As reported in Larry Thompson, *Correcting the Code: Inventing the Genetic Cure for the Human Body* (New York: Simon and Schuster, 1994), p. 311.

Meanwhile Ciba-Geigy. See "FTC Accord in Ciba Geigy/Sandoz Merger," Federal Trade Commission press release, Dec. 17, 1996.

About four thousand diseases are caused. Brown, "Major Pharmaceutical Companies Infuse Needed Capital."

49 *will top $2 billion.* Ibid.

will reach upward of $45 billion. "FTC Accord" press release.

$63 billion merger. Interview with Joseph Schepars, Novartis spokesperson, Sept. 1997.

"the company will be able to pick and choose." Glorioso quoted in Gavaghan, "NIH Wins Patent."

"only a tiny handful of entities." "FTC Accord" press release.

"Given the combination." Ibid.

51 *Anderson complains.* Anderson quoted in Ruth Sorelle, "The Gene Doctors," *Houston Chronicle,* special series, Houston Chronicle Publishing, 1995.

"In an era of decreased federal funding." Steven A. Rosenberg, "Secrecy in Medical Research," *FASEB Newsletter,* June 1996. See also Lawrence K. Altman, "Se-

crecy Is Hurting Medical Research, a U.S. Official Says," *New York Times*, February 10, 1996.

52 *now being denied a promising treatment approved by the Food and Drug Administration.* See "CellPro's 'Rick Project': Remove Cancer in 8 Weeks," *Seattle Times*, May 9, 1997. See also Elaine Lafferty, "By His Own Device," *Time*, May 19, 1997, p. 70; and Eliot Marshall, "Varmus to Rule in Fight over Cell-Sorting Technology," *Science*, June 6, 1997, p. 1488.

53 *"It's unfortunate that these sorts of things."* Yeager quoted in Keith Ervin, "Patent Litigation Threatens Cell-Therapy Progress," *Seattle Times*, Apr. 17, 1997.
made an appeal on CellPro's behalf. Letter from Lloyd N. Cutler and Birch Bayh, Jr., to Donna Shalala, Mar. 3, 1997, made available by CellPro, Inc.
"would pose a grave threat." Letter from Gerhard Casper to Harold Varmus, June 10, 1997, made available by the Center for the Study of Responsive Law.

54 *"Could you patent the sun?"* Salk quoted in Philip L. Bereano, "Patent Pending: The Race to Own DNA," part 2, *Seattle Times*, Aug. 27, 1995, p. B5.
another prominent case attracted. Kurt Eichenwald, "Push for Royalties Threatens Use of Down Syndrome Test," *New York Times*, May 23, 1997, p. 1. See also "Scientist Seeks Royalties for Down Syndrome Test," *USA Today*, Sept. 9, 1997.

55 *laboratories owned by SmithKline Beecham.* Interview with Andrew Dhuey, Nov. 1997. See also "Another hCG Patent Lawsuit Settled," *Clinical Laboratory News* (American Association for Clinical Chemistry), Sept. 1997.
"To claim private ownership rights." Interview with Arnold Relman, Dec. 1997.
"does seem a little like patenting." Merges quoted in Eichenwald, "Push for Royalties Threatens Use," p. 1.

56 *the HMO has contested the claim.* Interview with Mitchell Sugarman, Oct. 1997.
"a dangerous attack." Interview with Michael Watson, vice president, American College of Medical Genetics, Oct. 1997.
"Most people are really outraged." Demers quoted in Eichenwald, "Push for Royalties Threatens Use," p. 1.
"pivotal decision." Letter from Andrew J. Dhuey to "Laboratories Solicited by the American College of Medical Genetics," June 6, 1997.
he could earn as much as $100 million. Interview with Mitchell Sugarman, Oct. 1997.
"Why should they be allowed?" Bogart quoted in Eichenwald, "Push for Royalties Threatens Use," p. 1.

57 *"we will need to go back at this again."* Ibid.
"his financially motivated claim." Interview with Mitchell Sugarman, Nov. 1997. See also "Kaiser Permanente Files Suit," Kaiser Permanente press release, Aug. 28, 1997.
"free exchange of medical knowledge." See Jack A. Singer, "The Free Exchange of Medical and Surgical Knowledge," paper presented at the American Society of Cataract and Refractive Surgery Symposium, Boston, Apr. 10, 1994.

58 *"If this patent is enforced."* Evans quoted in "Down Syndrome Test Debated," ABC News, Sept. 9, 1997, transcript from ABCNews.com.

5. SOFTWARS

59 *sweeping ownership rights.* U.S. Patent No. 5,241,671, "Multimedia Search System Using a Plurality of Entry Path Means Which Indicate Interrelatedness of Information," Aug. 31, 1993. The patent was rescinded within a year. See also Peter Lewis, "The New Patent That Is Infuriating the Multimedia Industry," *New York Times,* Nov. 28, 1993.

"We invented multimedia." Frank quoted in Simson L. Garfinkel, "How Could the Patent Office Ever Grant a Patent to Compton's on Its Claim to Have Invented Multimedia? This Is How," *Wired,* July 1994, p. 109.

61 *"one of the most serious developments."* Dodds quoted in Audrey Doyle, "New Compton's Patent Sparks Controversy in Multimedia Circles," *Computer Graphics World,* Jan. 1994.

"a 41-count snow job." Lippincott quoted in Garfinkel, "How Could the Patent Office," p. 109.

"this sort of search system." Bastin quoted in Doyle, "New Compton's Patent."

explored originally at . . . Xerox PARC. See, for instance, Fred Warshofsky, *Patent Wars: The Battle to Own the World's Technology* (New York: John Wiley and Sons, 1994), chapter 3.

"Patenting multimedia." Carberry quoted in Clair Whitmer, "Compton's Multimedia Patent, Licensing Plan Worry Developers," *InfoWorld,* Nov. 29, 1993.

"like a herd of elephants." Min Yee quoted ibid.

Joe Clark captured the prevailing sentiment. United States Patent and Trademark Office, "Public Hearing on Use of the Patent System to Protect Software-Related Inventions," Jan. 26–27, 1994, San Jose, Calif., transcript of proceedings.

62 *software has proven exceptionally troublesome.* See, for example, Anthony Lawrence Clapes, *Softwars: The Legal Battles for Control of the Global Software Industry* (Westport, Conn.: Quorum Books), 1993.

63 *The Supreme Court ruled in 1972. Gottschalk v. Benson,* 409 U.S. 63 (1972).

64 *being issued at nearly three times.* My calculation, based on figures from the U.S. Patent Office and from *Patnews,* Internet patent news service, 1997.

unprecedented hearing in Silicon Valley. Unless otherwise cited, testimony is drawn from the transcript of "Public Hearing on Use of the Patent System," San Jose, Jan. 1994.

69 *Within a year of the Silicon Valley hearing.* See Elinor Mills, "Annulling Compton's Patent Helps PTO Earn Industry's Trust," *InfoWorld,* Nov. 14, 1994, p. 38.

70 *"Bhopal of software patents."* Mitch Kapor quoted in Gino Del Guercio, "Softwars," *World Monitor,* Oct. 1991, p. 22.

the case of Vern Blanchard. United States Patent and Trademark Office, "Public Hearing on Use of the Patent System to Protect Software-Related Inventions," Feb. 10–11, 1994, Arlington, Va., transcript of proceedings.

XyWrite word processing program. See Simson L. Garfinkel, Richard M. Stallman, and Mitchell Kapor, "Why Patents Are Bad for Software," in Peter Ludlow, ed., *High Noon on the Electronic Frontier* (Cambridge, Mass.: MIT Press, 1996), pp. 35–47.

71 *"More than half of the speakers."* "Public Hearing on Use of the Patent System," San Jose, Jan. 1994.

A 1997 study. Steve Lohr, "Study Ranks Software as No. 3 Industry," *New York Times,* June 3, 1997. The study was conducted by Nathan Associates, an economic consulting firm in Arlington, Va.

11,500 software patents. Gregory Aharonian, "A Mess of Recent Bad Software Patent Claims," *Patnews,* July 5, 1997.

72 *patent no. 4,648,067.* "Footnote Management for Display and Printing," issued Mar. 1987, assigned to IBM. See also Brian Kahin, "The Software Patent Crisis," *Technology Review,* Apr. 1990, p. 52.

patent no. 4,807,182. "Apparatus and Method for Comparing Data Groups," issued Feb. 1989, assigned to Advanced Software, Inc.

patent no. 5,642,503. "Method and Computer System for Implementing Concurrent Accesses of a Database Record by Multiple Users," June 24, 1997, awarded to Microsoft.

patent no. 5,642,430. "Visual Presentation System Which Determines Length of Time to Present Each Slide or Transparency," June 24, 1997, assigned to IBM.

Xerox even won a patent. U.S. Patent No. 5,638,543, "Method and Apparatus for Automatic Document Summarization," awarded June 10, 1997.

Merrill Lynch patented. See Richard H. Zaitlen and Wendy Young, "Software Patents: The Only Game in Town?" *MicroTimes,* May 1996.

"the Crystal City test." Interview with Greg Aharonian, Sept. 1997. See also Greg Aharonian, "40,000 Questionable Software Patents by the End of the Decade," *Patnews,* June 24, 1997.

73 *A visit to the U.S. Patent and Trademark headquarters.* Unless otherwise noted, facts derive from my visit in Mar. 1995.

74 *report of a 1966 presidential commission.* "To Promote the Progress of Useful Arts," Report of the President's Commission on the Patent System, 1966, as quoted in Aharonian, "40,000 Questionable Software Patents."

"In this field especially." Interview with Alan MacDonald, Mar. 1995.

75 *"tremendous problem retaining examiners."* Interview with Charles Van Horn, Mar. 1995.

Brunelleschi claimed that he had invented. See Steven Saas, "Brunelleschi's Bargain: Intellectual Property in Digital Space," in Federal Reserve Bank of Boston, *Regional Review,* Fall 1993, p. 6.

76 *"A biotechnology start-up firm."* Interview with Robert Merges, Sept. 1995.

battle between Microsoft and Sun. See John Markoff, "Sun Sues Microsoft on Use of Java System," *New York Times,* Oct. 8, 1997.

77 *according to John Heilemann.* John Heilemann, "The Sun King: Letter from Silicon Valley," *New Yorker,* Mar. 16, 1998, pp. 30–35.

77 *Microsoft's "exclusionary business practices."* McNealy's testimony quoted ibid.
"a horrible disaster." Interview with Richard Stallman, Mar. 1998.

78 *"The expensive patent process."* "Public Hearing on Use of the Patent System," San Jose, Jan. 1994.
at least fifty-two cross-licensing agreements. Mo Krochmal, "IBM Collects on Intellectual Property," *Tech Web News,* Nov. 6, 1997. See also "IBM and Novell Sign Patent Cross-Licensing Agreement," *Business Wire,* Nov. 6, 1997.
"sue-the-bastards approach." Norm Alster, "New Profits from Patents," *Fortune,* Apr. 25, 1988, pp. 185–90.
company called E-data. See Neil Gross and Amy Cortese, "E-Commerce: Who Owns the Rights?" *Business Week,* July 29, 1996, p. 65. "E-Data Files 40 Patent Infringement Claims against Thirteen Defendants in New York Court," press release, Interactive Gift Express Inc., Aug. 26, 1996.
75,000 separate companies. See Shoshana Berger, "Patently Offensive," *Wired,* vol. 4.09, on-line@hotwired.com.
a carrot and a stick. See Keith Dawson, "A Carrot, a Stick, and a Judge," *Tasty Bits from the Technology Front,* on-line newsletter, July 14, 1996.

79 *Charles Freeny won a patent.* See "The True-Life Drama of the E-data System Patent," backgrounder, E-data Corporation.
"pit bull of patent infringement." See Robert Metz, "Fallen Angel," Shaking the Money Tree column, *Money Talks* (daily investment magazine), Nov. 6, 1995.
"three men and a patent." Ibid.

80 *stock price soared.* See, for example, Robert Gebeloff, "Internet Commerce Patented?" *Bergen (N.J.) Record,* Sept. 16, 1997.
lawsuits against forty-one companies. Time line and supporting information about the case published on-line by defense counsel, Oppedahl and Larson, at the Oppedahl & Larson Patent Law home page, http://www.patents.com/ige.sht.
"Part of our marketing strategy." Freilich quoted in Greg Aharonian, "How Much Bad Software Patents Will Cost You," *Patnews,* May 26, 1996.
"We have clients." Baker quoted in Noah Robischon, "Patent Spending," *Netly News,* on-line newsletter, June 6, 1996.
defendants seek to declaw. See, for example, *Interactive Gift Express Incorporated v. CompuServe Incorporated et al.,* United States District Court for the Southern District of New York, Civil Action No. 95 CIV. 6871 (BSJ). See also "CompuServe Applauds Federal Judge Court Order," press release, CompuServe, Inc., July 3, 1996.

81 *"Two kinds of patents."* Interview with Richard Stallman, Mar. 1998. See also League for Programming Freedom, "Against Software Patents," in Ludlow, ed., *High Noon on the Electronic Frontier,* pp. 47–63.
software firm called Open Market. Keith Dawson, "Open Market Awarded Three Basic Patents," *Tasty Bits from the Technology Front,* Mar. 9, 1998.

82 *"There will be no easy knockout."* Interview with Greg Aharonian, Mar. 1998.
"Let them all sue." Ibid.

6. SOYBEAN DREAMS

83 *As Becky remembers it.* Interview with Becky Winterboer, Apr. 1997.
"We had no idea." Interview with Denny Winterboer, Apr. 1997.
The Winterboers' crime. Paul M. Barrett, "High-Court Battle Sprouts from Clash between Farmers and the Seed Industry," *Wall Street Journal,* May 23, 1994, p. B1.
brown-bag sales. Hope Shand, "U.S. Congress Restricts Farmers' Rights," *Seedling,* Oct. 1994.

84 *The company claimed proprietary rights. Asgrow Seed Co. v. Winterboer,* U.S. Supreme Court no. 92-2038, Jan. 18, 1995.
"farmers helping farmers." Barrett, "High-Court Battle," p. B1.

85 *Among the benefits of the scheme.* See Jack Doyle, *Altered Harvest: Agriculture, Genetics and the Fate of the World's Food Supply* (New York: Viking Penguin, 1985), p. 164.
"a packet of genetic information." Jack R. Kloppenburg, Jr., *First the Seed: The Political Economy of Plant Biotechnology* (Cambridge: Cambridge University Press, 1988), p. 201. See also Seth Shulman, "Seeds of Controversy," *BioScience* 36, no. 10 (Nov. 1986).
"a developer, producer and marketer of genetics." Asgrow Seed Company, Asgrow Seed Technology Web Site, 1997.

86 *"The major food plants."* Wilkes quoted in A. B. Cunningham, "Indigenous Knowledge and Biodiversity," *Cultural Survival Quarterly* 15, no. 3 (Summer 1991).

87 *The crop did originate in China.* "Soybeans . . . the Miracle Crop," *Soy Stats,* American Soybean Association, 1996.
According to a U.S. government estimate. As cited in Hope Shand's testimony before the U.S. House of Representatives, Committee on Agriculture, Subcommittee on Department Operations and Nutrition, May 24, 1994.
"a man can patent a mousetrap." Burbank quoted in Kloppenburg, *First the Seed,* p. 130.

88 *one of the department's central mandates.* Ibid., p. 59.
even used as ballast. Soy Stats, 1996.
"laid the foundation." Ibid.
"private control over the gene pool." Interview with Denny Winterboer, Apr. 1997.
some five hundred thousand farmers in India. See Chirag Mehta, "The Seed Satyagraha: Indian Farmers and Global Capital Face Off," *Dollars and Sense,* Sept.–Oct. 1994, p. 25.

89 *"the issue of control of seed varieties."* Shand's testimony, May 24, 1994.
filed an amicus brief. Asgrow Seed Co. v. Winterboer, Supreme Court No. 92-2038, Oct. term, 1994.

89 *"Congress wanted to allow."* Justice John Paul Stevens in his dissenting opinion, *Asgrow Seed Co. v. Winterboer.*

90 *"a crushing blow to farmers' rights."* Shand, "U.S. Congress Restricts Farmers' Rights."

dismiss as a paranoid fantasy. Interview with Hope Shand, Apr. 1997.

91 *companies dominate the global seed trade.* Industry information compiled by Inverizon International, Inc., business consultants, St. Louis, Mo., 1996. See also "Multinational Seed Corp. Consolidations a Threat to Ag Seed Diversity and Genetics," FedCo. seed catalog, 1997.

Agracetus won a 1994 European patent. See Seth Shulman, "Patent Medicine," *Technology Review,* Nov.–Dec. 1995, pp. 29–36.

"all crops will be transgenic." Interview with Russell Smestad, former vice president, Agracetus, Inc., Apr. 1994.

92 *"at a stroke of a pen."* Hawtin quoted in "Species Patent on Transgenic Soybeans," *RAFI Communiqué,* Mar.–Apr. 1994.

the company may even realize its goal. Ibid.

Ning-Sun Yang. Jennifer Galloway, "Gene Gun Goes into Clinical Trials," *Wisconsin State Journal,* Oct. 22, 1996. See also "The Subtle Approach," *Discover,* special issue, *The Year in Science,* Jan. 1996.

93 *Agracetus was a biotechnology start-up.* Background information provided by company officials in 1994 and by Ning-Sun Yang, May 1997.

bullets would be the size of a tennis ball. Interview with Ning-Sun Yang, May 1997.

94 *"sitting on a gold mine."* Ibid.

after reading a pathbreaking article. T. Klein, Edward Wolf, et al., "High-Velocity Microprojectiles for Delivering Nucleic Acids into Living Cells," *Nature* 327 (1987): 70–73.

Agracetus made some important contributions. See, for example, Ning-Sun Yang, Carolyn De Luna, and Liang Cheng, "Gene Transfer via Particle Bombardment: Applications of the Accell Gene Gun," in Jon A. Wolff, ed., *Gene Therapeutics: Methods and Applications of Direct Gene Transfer* (Boston: Birkhauser, 1994), pp. 193–208.

"particle bombardment technology." U.S. Patent No. 4,945,050, "Method for Transporting Substances into Living Cells and Tissues and Apparatus Therefor," issued to John C. Sanford, Edward D. Wolf, and Nelson K. Allen, July 31, 1990, assigned to the Cornell Research Foundation, Inc.

95 *"Using the particle acceleration method."* See Friends of the Earth Europe, "Boycott of Genetically Modified Soya Gains Momentum," *Biotechnology Programme,* Nov. 1, 1996.

"no one knew how to get a gene into a plant." Smestad quoted in Karen Bernstein, "Contemplating Agbiotech's Future," *BioCentury,* Dec. 4, 1995.

a kind of contract farming. Ibid.

96 *"There are many battles."* Interview with Hope Shand, Mar. 1997.

97 *"shut our work down."* Interview with Melvin Oliver, Mar. 1994.
step of legally contesting the patent. See Seth Shulman, "A New King Cotton," *Technology Review,* July 16, 1994.
"adverse impact." Interviews with Howard Silverstein, Mar. 1994 and Mar. 1997.
"We question the very foundation." Interview with Hope Shand, May 1997.
"just like Polaroid." Wrage quoted in Hope Shand, "Control of Cotton: The Patenting of Transgenic Cotton," *RAFI Communiqué,* July–Aug. 1993, p. 4.
a $1 billion settlement. See Fred Warshofsky, *Patent Wars: The Battle to Own the World's Technology* (New York: John Wiley and Sons, 1994), p. 87.

98 *agricultural breakthrough: hybrid crops.* Kloppenburg, *First the Seed,* pp. 123–32; see also Frederick H. Buttel and Jill Belsky, "Biotechnology, Plant Breeding, and Intellectual Property: Social and Ethical Dimensions," in Weil and Snapper, *Owning Scientific and Technical Information,* pp. 120–22.
Diamond v. Chakrabarty. 447 U.S. 305 110 S. Ct. 2204 and 447 U.S. 303, 308–9 (1980).
Ananda M. Chakrabarty. See Doreen Stabinsky, "Who Owns Life: A Short History of Plant Patents," *GeneWatch* 10, no. 2–3 (Oct. 1996).

99 *"anything under the sun."* Supreme Court decision, *Diamond v. Chakrabarty.*
"the element of design resides in nature." See Mark Sagoff, "Animals as Inventions: Biotechnology and Intellectual Property Rights," *Institute for Philosophy and Public Policy* 16, no. 1 (Winter 1996).

100 *"analogies are so easy to draw."* Interview with John Barton, July 1995; see also John H. Barton, "Patenting Life," *Scientific American* 264 (Mar., 1991): 40.
patent on a "transgenic mouse." Keith Schneider, "Harvard Wins First Animal Patent — for Building a Better Mouse," *New York Times,* Apr. 17, 1988, p. E5. See also Keith Schneider, "Patenting Life: U.S. Policy Decision Sets Off a Debate over Morals and Lure of Vast Profits," *New York Times,* Apr. 18, 1987. (The U.S. Patent Office opened the door for animal patenting in April 1987, when it announced a policy shift. The first patent on an animal variety was granted a year later.)
rescinded the cotton patent. See Fred Powledge, "Who Owns Rice and Beans? Patents on Plant Germplasm," *BioScience,* July 1995.

101 *the odd alliance.* Teresa Riordan, "U.S. Revokes Cotton Patents after Outcry from Industry," *New York Times,* Dec. 8, 1996, p. D1.
a $150 million deal. Press release, Monsanto Company, Apr. 8, 1996.

102 *uncomfortable silence.* Interview with Karen K. Marshall, Monsanto spokeswoman, May 1997. See also "Monsanto's About-Face on Agracetus' Soybean Species Patent," *RAFI Communiqué,* July–Aug. 1996.
"if you've met one farmer." Unless otherwise noted, quotes are from an interview with John McClendon, July 1997.

103 *"technology fee."* Verlyn Klinkenborg, "Biotechnology and the Future of Agriculture," *New York Times,* Dec. 8, 1997, p. A22.
Roundup Ready Gene Agreement. See "Bioserfdom: Technology, Intellectual

103 Property and the Erosion of Farmers' Rights in the Industrialized World," *RAFI Communiqué,* Mar.–Apr. 1997.
have proved a dismal failure. Klinkenborg, "Biotechnology and the Future," p. A22.

104 *once boasted some 2 million farmers.* Farming industry figures from McClendon. See also Inverizon, statistics from St. Louis, 1996.

7. TABORSKY'S LAMENT

106 *His tragic tale.* This section derives primarily from reporting done in 1996, including interviews with all parties to the case. See Seth Shulman, "A Researcher's Conviction," *Technology Review,* Jan. 1997, p. 20.
U.S. Patent No. 5,082,813. "Aluminosilicates with Modified Cation Affinity," issued to Petr Taborsky Jan. 21, 1992.
"It's remarkable." Lange quoted in Mireya Navarro, "Dispute Turns a Researcher into an Inmate," *New York Times,* June 9, 1996, p. 22.

107 *a job earning $8.50.* "Disputes Rise over Intellectual Property Rights," "Morning Edition," National Public Radio, Sept. 30, 1996, including interviews with Petr Taborsky and University of South Florida attorney Noreen Segress.
Florida Progress had paid the university. See William Booth, "From University Lab to the Chain Gang," *Washington Post,* June 17, 1996, p. A1.

108 *Taborsky had shown promise.* See "Taborsky May Soon Move from Prison," *Bradeton Herald,* Nov. 1996.
"My life has been made miserable." Taborsky quoted in Rebecca Gilbert, "Ex-Student Offers USF Compromise in Patent Suit," *Oracle News Service,* Nov. 1996.
Taborsky had a breakthrough. See Shulman, "A Researcher's Conviction," p. 20.

109 *"wasn't going to let them intimidate me."* Taborsky quoted in Lisa Holewa, "Chemistry Student Sent to Chain Gang over Patent Dispute," Associated Press, Oct. 17, 1996.
According to a police report. Ibid.

110 *his marriage crumbled.* Ibid.
a three-and-a-half-year sentence. Rebecca Gilbert, "Report Examines USF's Role in Patent Suit," *Oracle News Service,* Oct. 17, 1996, p. 2.
"There are a lot of things." Interview with Dexter Douglass, Nov. 1996.
"wrongful conviction." Gilbert, "Report Examines USF's Role," p. 2.
tangle of civil legal actions. Interview with Henry Lavendera, Mar. 1998.
actions of Francis Borkowski. Interview with Henry Lavendera, Nov. 1996.

111 *"We are concerned."* Ibid.
University of South Florida. University statistics from Navarro, "Dispute Turns a Researcher into an Inmate," p. 22.
roughly $1.5 billion. Figures provided by the National Science Foundation press office. See Shulman, "A Researcher's Conviction," p. 20.
"This case is just the tip." Reichman quoted in Booth, "From University Lab to Chain Gang," p. A1.

112 *the "bizarre" nature.* Interview with Cornelius Pings, Oct. 1996.

they "are the foundation." Address to the National Press Club, Washington, D.C., July 18, 1995, reprinted in *Technology Review* (MIT alumni edition), Oct. 1995, p. 11.

graduates have founded 4,000 firms. From "MIT: The Impact of Innovation," report of a study sponsored by BankBoston, Mar. 5, 1997. See "Study Reveals Major Impact of Companies Started by MIT Alums," *MIT Tech Talk,* Mar. 5, 1997, p. 1.

113 *twenty-fourth largest economy.* Ibid.

David Kern claims. See Wade Roush, "Secrecy Dispute Pits Brown Researcher against Company," *Science,* Apr. 25, 1997, p. 523.

a battle arose in 1995 over software. Jane Baird, "Stakes Huge in Legal War for Software Codes," *Houston Chronicle,* July 25, 1995. See also Todd Ackerman, "Research Firm Still Struggling for the Home Run after 15 Years," *Houston Chronicle,* Nov. 28, 1997.

stronger protection for scientists. See Gina Kolata, "Safeguards Urged for Researchers: Aim Is to Keep Vested Interests from Suppressing Discoveries," *New York Times,* Apr. 17, 1997, p. A24.

114 *One 1997 study examined.* David Blumenthal et al., "Relationships between Academic Institutions and Industry in the Life Sciences — an Industry Survey," *New England Journal of Medicine,* Feb. 8, 1996, p. 368.

Another set of researchers. See Sheldon Krimsky et al., "Financial Interest of Authors in Scientific Journals: A Pilot Study of 14 Publications," *Science and Engineering Ethics* 2, no. 4 (1996): 395–410.

the Bayh-Dole Act. For a discussion of the legislative history, see Linda Marsa, *Prescription for Profits* (New York: Scribners, 1997) p. 96.

"selling the tree of knowledge." Gore quoted in David Noble, "Academia Incorporated," *Science for the People,* Jan./Feb. 1983, p. 7.

115 *"Based on forty years' experience."* Rickover quoted in Marsa, *Prescription for Profits,* pp. 97–98.

found secrecy particularly widespread. As reported by Eliot Marshall, "Secretiveness Found Widespread in Life Sciences," *Science,* Apr. 25, 1997, p. 525.

material transfer agreements. This discussion and the comments of Julie Norris and Keith Yamamoto draw from Eliot Marshall, "Need a Reagent? Just Sign Here . . ." *Science,* Oct. 10, 1997, p. 212. See also Steven Benowitz, "Is Corporate Research Funding Leading to Secrecy in Science?" *The Scientist,* Apr. 1, 1996, p. 1.

116 *"We sit here and talk."* Berg quoted in Colin Macilwain, "Conflict-of-Interest Debate Stirs Mixed Reaction at NIH," *Nature,* Feb. 3, 1994, p. 401.

"Today's fundamental elucidation of a protein." See "Values Poisoned by Commerce," *Nature,* Feb. 15, 1996, p. 567.

John Kenneth Galbraith. See *The Good Society: The Humane Agenda* (Boston: Houghton Mifflin, 1996), p. 97.

117 *Harvard Medical Science Partners.* For a discussion of the arrangement, see Jaron Bourke and Robert Weissman, "The Entrepreneurial University," *Academe,* Sept.–Oct. 1990, p. 15. See also Seth Shulman, "Academic, Inc." *Technology Review,* Nov./Dec. 1987, pp. 11–12.

"the basic problem with the Harvard fund." Leahey quoted in Jaron Bourke and Robert Weissman, "The Entrepreneurial University," p. 15.

"it will take very strong leadership." Bok quoted in Susannah Hunnewell, "The Medical-Industrial Complex," *Harvard Magazine,* Jan.–Feb. 1994, p. 37.

Sir John Maddox. See John Maddox, "Can the Research University Survive?" *Nature,* June 30, 1994, p. 703.

118 *"washed up, kaput, dead, gone."* Bernard Margolis quoted in John Yemma, "Book Value," *Boston Globe Magazine,* Mar. 29, 1998, p. 25.

some 16,000 public library branches. Information sheets provided by the American Library Association, 1997.

"We hate to take information resources." Interview with Alan Allaire, Mar. 1998.

the case of Lloyd Davidson. See Seth Shulman, "Pay-per-View Libraries," *Technology Review,* Oct. 1992, p. 14.

119 *"Proprietary information."* Ibid. Based on interview with Karen Muller, July 1992.

"those who make the investments." See "Database Protection — the Time Is Now," Information Industry Association report, 1997.

"fair-use" provisions. For an in-depth discussion, see Neil Weinstock Netanel, "Copyright and a Democratic Civil Society," *Yale Law Journal,* Nov. 1996, pp. 283–387.

120 *"As much as I hate to see it happen."* Romano quoted in Yemma, "Book Value," p. 26.

right to publish residents' phone numbers. See the Supreme Court decision in *Feist Publications, Inc. v. Rural Telephone Service Co.,* 111 S. Ct. 12282 (1991).

overruled the West Publishing Company. See David Cay Johnston, "West Publishing Loses a Decision on Copyright," *New York Times,* May 21, 1997, p. D1.

these decisions represent a threat. See "Database Protection — the Time Is Now."

121 *"There's a technical legal term."* Interview with Adam Eisgrau, Jan. 1998.

"The full potential." Working Group on Intellectual Property Rights, *Intellectual Property and the National Information Infrastructure* (Washington, D.C.: Information Infrastructure Task Force), Sept. 1995, p. 10.

end run around Congress. Wendy Lubetkin, "U.S. Says 'Fair Use' Should Apply on the Internet," *USIS Geneva Daily Bulletin,* Dec. 13, 1996. See also "Statement from U.S. Delegation to WIPO Conference," *USIS Geneva Daily Bulletin,* Dec. 13, 1996. And see Pamela Samuelson, "Copyright and Digital Libraries," *Communications of the ACM* (Association for Computing Machinery), Apr. 1995, pp. 15–23.

122 *a flurry of criticism.* See, for example, Andrew Lawler, "Treaty Draft Raises Scientific Hackles," *Science,* Oct. 25, 1996, p. 494.

"antithetical to the principle." Quoted in Robert M. White, "Taking on the

122 Database Challenge and Winning," *Technology Review*, May–June 1997.
"We can't take the future of libraries." Interview with Adam Eisgrau, Feb. 1998.
National Library of Medicine. Press release, National Institutes of Health, "Library of Medicine Database Opened to the Public," June 30, 1997. Access on-line at http://www.ncbi.nlm.nih.gov/PubMed.

123 *"There is no good market reason."* Robert Kuttner, *Everything for Sale* (New York: Knopf, 1997), pp. 199–200.
"Visit your local library." Clifford Stoll, *Silicon Snake Oil* (New York: Doubleday, 1995), p. 207.
"Libraries in America are situated." Peter Lyman, "What Is a Digital Library? Technology, Intellectual Property, and the Public Interest," *Daedalus*, special issue, *Books, Bricks, and Bytes*, Fall 1996, p. 3.
museums also face a barrage. See Seth Shulman, "Digital Museums," *Technology Review*, Nov./Dec. 1994, pp. 20–21.
"It's open season." Interview with Janice Sarkow, Aug. 1994.

124 *the privately held Corbis Corporation.* Information provided by Corbis Corp., Bellevue, Washington.
1.3 million images. See "Industry Leaders Join Corbis to Meet Growing Demand for Visual Content," *Business Wire*, Sept. 9, 1997.
"digital Alexandria." Carey Goldberg, "What's Wrong with This Picture?" *New York Times Magazine*, May 18, 1997, p. 23.
"comprehensive digital-image archive." Interview with Scott Sedlick, then marketing manager at Corbis, Aug. 1994.

125 *Art Museum Image Consortium.* See J. Trant and D. Berman, "The Art Museum Image Consortium: Licensing Museum Digital Documentation for Educational Use," *Spectra*, Fall 1997.

126 *"Until recently, interest in technology."* Interview with Katherine Jones Garmil, Mar. 1993.

8. PANNING FOR DRUGS

127 *U.S. Plant Patent 5,751.* "Banisteriopsis caapi (cv) 'Da Vine,'" issued to Loren S. Miller, June 17, 1986, claiming "A new and distinct Banisteriopsis caapi plant . . . and its medicinal properties."
an intellectual property treaty. An initial agreement with the United States was signed in 1993 by President Sixto Duran Ballen, but three years later the parliament rejected the treaty. See Mario Gonzalez, "Ecuador: Controversial Patent Agreement with U.S. Up for Revision," Inter Press Service, Aug. 7, 1997.
some four hundred indigenous groups. See Julie Watson, "Amazon Tribes Fight Patent on Sacred Vine," *Dallas Morning News*, Aug. 28, 1996.
"a true affront." Coordinating Body of Indigenous Organizations of the Amazon Basin (COICA), press release, Quito, Ecuador, June 26, 1996. See also Josep M. Fericgla, "Ayahuasca Patented!" *Eleusis* (bulletin of the Italian Society for the Study of States of Consciousness), no. 5, 1996, pp. 19–20.

128 *sales in excess of $250 billion.* See Greg Critser, "Oh, How Happy We Will Be: Pills, Paradise, and the Profits of the Drug Companies," *Harper's,* June 1996, p. 39.

129 *analgesic properties of willow bark.* See James Glanz, "Industry Does a Double-Take on Plant-Based Drugs," *R&D,* July 1993, p. 41.

streptomycin was developed. Press material supplied by Merck and Co., Feb. 1997.

"I think people are really waking up." Hylands quoted in Glanz, "Industry Does a Double-Take," p. 41.

Miller expresses only bitterness. Interview with Loren Miller, Oct. 1997.

130 *The Yage Letters.* Alexander Cockburn, "Bill Burroughs and the Biopirates," *Nation,* Aug. 25–Sept. 1, 1997, p. 8. See also Ed Dravo, "Java Jive: Evaluating the Stocks of Starbucks and Shaman Pharmaceuticals," *Financial World,* Aug. 15, 1994, p. 78.

Schultes's collection. Information provided by the Harvard Herbaria, Cambridge, Mass.

Pharmaceutical firms learned. See Gail Fowler, "Finding Medicines in Nature's Drug Store," *MSD World/Merck World* (internal corporate newsletter), Oct. 1997, p. 3.

132 *"Without a robust patent system."* Interview with Patricia Granados, Dec. 1997.

"drug companies now spend $250 million." Interview with Charles Van Horn, Mar. 1996. Some contend that the average cost to bring a drug to market is as much as $359 million. See, for example, Gerald J. Mossinghoff and Thomas Bombelles, "The Importance of Intellectual Property Protection to the American Research-Intense Pharmaceutical Industry," *Columbia Journal of World Business,* Spring 1996, p. 38.

Congress passed the Plant Patent Act. Jack R. Kloppenburg, Jr., *First the Seed: The Political Economy of Plant Biotechnology* (Cambridge: Cambridge University Press, 1988), pp. 132–33. "Sixty-Five Years of the U.S. Plant Patent Act (PPA)," *RAFI Communiqué,* Nov.–Dec. 1995.

133 *earn the pharmaceutical industry some $32 billion.* See *Enclosures of the Mind: Intellectual Monopolies: A Resource Kit on Community Knowledge, Biodiversity and Intellectual Property* (Ottawa: Community Biodiversity Development and Conservation Program, Rural Advancement Foundation International [RAFI], 1996), p. 6.

multinational corporations held 79 percent. "Utility Plant Patents: A Review of the U.S. Experience, 1985–1995," *RAFI Communiqué,* July–Aug. 1995, p. 8.

Pharmagenesis. See Glanz, "Industry Does a Double-Take," p. 42.

Shaman Pharmaceuticals. Richard Phalon, "Keep Your Eye on the Ball," *Forbes,* Apr. 11, 1994, p. 78.

the drug curare. See Kathleen McAuliffe, "Shaman Pharmaceuticals," *Omni,* July 1993, p. 16.

bark of the South American tree. Alessandra Dalevi, "Green Piracy: Medicinal Herbs and the Plundering of the Amazon," *Brazil,* July 1997.

133 *MGI Pharma developed Salagen.* See "Oral Pilocarpine Hydrochloride for Radiation-Induced Dry Mouth," *Medical Sciences Bulletin,* Pharmaceutical Information Associates, June 1994. See also Dalevi, "Green Piracy."

134 *United Nations–sponsored report.* See Stanley Meisler, "UN: Drug Firms Exploit 3rd World," *Los Angeles Times,* Sept. 16, 1995, p. A2.

earn pharmaceutical firms some $5.4 billion. Dalevi, "Green Piracy."

"It's a question." Clay quoted in Daniel Goleman, "Shamans and Their Lore May Vanish with the Forest," *New York Times,* June 11, 1991, p. C1.

135 *an act of biopiracy.* Julie Watson, "Amazon Tribes Fight Patent on Sacred Vine," *Dallas Morning News,* Aug. 28, 1996.

even briefly occupied Ecuador's legislature. See "Biopiracy and Indigenous Knowledge," *Year-End Roundup,* Rural Advancement Fund International, Dec. 1996.

"The bilateral agreement." Cobo quoted in the Quito daily newspaper *Hoy.* For Cobo's statement and Grefa's response, see Gonzalez, "Ecuador: Controversial Patent Agreement."

136 *That country's rain forest.* For facts and figures on Costa Rica, see David Tenenbaum, "The Greening of Costa Rica," *Technology Review,* Oct. 1995, p. 42. See Christopher Joyce, *Earthly Goods: Medicine-Hunting in the Rainforest* (Boston: Little, Brown, 1994), pp. 122–27. See also "Good Chemistry: Thomas Eisner's Nose for Knowledge," *Amicus Journal,* Spring 1993, p. 16.

rapid disappearance of species. For a discussion of extinction, see Fogarty International Center, "Report of a Special Panel of Experts on the International Cooperative Biodiversity Groups (ICBG)," Bethesda, Md., Aug. 1997.

he wrote a landmark paper. Thomas Eisner, "Prospecting for Nature's Chemical Riches," *Issues in Science and Technology,* Winter 1989. See also "Chemical Prospecting: A Proposal for Action," reprinted in F. H. Bormann and S. R. Kellert, eds., *Ecology, Economics, Ethics: The Broken Circle* (New Haven: Yale University Press, 1991), p. 196.

137 *In the 1991 deal.* Merck press releases, "Merck & Co., Inc. and Costa Rica's National Biodiversity Institute Renew Innovative Collaborative Agreement to Search for New Drugs in Biological Samples," Feb. 4, 1997; Jan. 28, 1997; Dec. 5, 1996; July 28, 1994; and "INBio of Costa Rica and Merck Enter into Innovative Agreement to Collect Biological Samples while Protecting Rain Forest," Sept. 19, 1991. See also Christopher Joyce, "Prospectors for Tropical Medicines," *New Scientist,* Oct. 19, 1991, pp. 36–40; J. Preston, "A Biodiversity Pact with a Premium," *Washington Post,* June 9, 1992, p. A16.

Merck sold nearly $20 billion. Merck year-end financial data, Dec. 31, 1996; quarterly financial report, Mar. 31, 1997.

Merck's three top-selling drugs. Stephen S. Hall, "Success Is Like a Drug," *New York Times Magazine,* Nov. 23, 1997, p. 69.

138 *Landowners in Costa Rica are often interested.* Elissa Blum, "Making Biodiversity Conservation Profitable: A Case Study of the Merck/INBio Agreement," *Environment,* May 1993, pp. 16–29.

"the most efficient way." Mathews quoted ibid.

138 *the deal has yet to result in any actual drugs.* Interview with John Doorley, Jan. 1998.

139 *counted some fifty-one companies.* See "List of Bioprospectors and Biopirates," *RAFI Communiqué,* Sept.–Oct. 1995.

pharmaceutical firm Diversa. "Conservation of Hot Spring Biodiversity: 21st-Century Priority for Yellowstone National Park," press release, World Foundation for Environment and Development, Aug. 17, 1997.

the agreement represents the shape. Varley quoted in "Yellowstone 125th Birthday Celebration Tainted," press release, Edmonds Institute, Aug. 15, 1997.

a legal challenge from two U.S. nonprofit groups. "Yellowstone Deal's Sleight of Hand: Company Wants Genes, not Biodiversity," *RAFI Communiqué,* Sept.–Oct. 1997.

140 *The issue of ownership.* Blum, "Making Biodiversity Conservation Profitable."

"the equitable sharing of benefits." The Convention on Biological Diversity, commonly referred to as the Biodiversity Treaty, was one of two major treaties open for signature at the United Nations Conference on Environment and Development (UNCED) in 1992. Having secured its thirtieth ratification in September 1993, the treaty entered into force Dec. 29, 1993. Benefit sharing is discussed in Article 8(j).

Trade Related Aspects of Intellectual Property (TRIPs). The industry perspective is given in Mossinghoff and Bombelles, "The Importance of Intellectual Property Protection."

141 *"just because the American Indian is there."* Bale quoted in Meisler, "UN: Drug Firms Exploit 3rd World."

they are not *inventions.* Sagoff quoted in Gary Taubes, "Scientists Attacked for 'Patenting' Pacific Tribe," *Science,* Nov. 17, 1995. See also Mark Sagoff, "Animals as Inventions: Biotechnology and Intellectual Property Rights," *Institute for Philosophy and Public Policy* 16, no. 1 (Winter 1996).

"What we did is show." Joyce, *Earthly Goods,* p. 127.

142 *"They are a closed club."* Ibid., p. 138.

Elaine Elisabetsky, a Brazilian pharmacologist. Elaine Elisabetsky, "Folklore, Tradition, or Know-How?" *Cultural Survival Quarterly,* Summer 1991, pp. 9–13.

143 *the case of Walter Lewis.* Interview with Edward Hammond, Oct. 1997. See also Fogarty International Center, "Report of a Special Panel of Experts."

International Cooperative Biodiversity Groups. For an overview of the program, see Fogarty International Center, "International Cooperative Biodiversity Groups: Overview," Biodiversity Program, Fogarty International Center, National Institutes of Health, Bethesda, Md., Aug. 1997.

"We energetically reject the lack of transparency." "Peru: Indigenous People 'Just Say No' to Bioprospector," *RAFI Communiqué,* Sept.– Oct. 1995, citing appeal by Peru's Aguaruna and Huambisa Council (CAH) to the U.S. National Institutes of Health, Mar. 10, 1995.

144 *patent on the use of turmeric.* U.S. Patent No. 5,401,504, "Use of Turmeric in Wound Healing," assigned to University of Mississippi Medical Center (Jack-

son, Miss.), filed Dec. 28, 1993. See "Herb Patent Withdrawn," *Nature*, Sept. 4, 1997.

144 *prior-art rules of the U.S. patent code.* Gregory Aharonian, *Patnews*, Internet patent news service, Nov. 15, 1996, responding to a report in *New Scientist*, Oct. 26, 1996.

145 *"In this day and age."* Ibid.

"inventors" are now seeking a narrower claim. "Turmeric Patent Overturned, but 'Inventors' Fight to Retain Claim," *RAFI Communiqué*, Sept.–Oct. 1997.

"knowledge is accumulated." Cox quoted in Mary Roach, "Secrets of the Shamans," *Discover*, Nov. 1993, p. 58.

"author-focused regime." James Boyle, *Shamans, Software, and Spleens: Law and the Construction of the Information Society* (Cambridge, Mass.: Harvard University Press, 1996), pp. 119–43.

146 *"The blindness of an author-centered regime."* Boyle, *Shamans, Software, and Spleens*, p. 130.

"patenting plants is no different." Lambert quoted in Fred Powledge, "Who Owns Rice and Beans?" *BioScience*, July–Aug. 1995, p. 440.

147 *a kind of layer-cake effect.* Eliot Marshall, "Snipping Away at Genome Patenting," *Science*, Sept. 19, 1997, pp. 1753–54.

"test case for the bio-age." "U.S. Patent on Tribesman's Blood Raises Ethical Questions," Associated Press, Goroka, Papua New Guinea, Apr. 20, 1996.

people known as the Hagahai. Taubes, "Scientists Attacked." See also "Indigenous Person's Cells Patented," *RAFI Communiqué*, Mar.–Apr. 1996.

148 *As Jenkins has explained.* Taubes, "Scientists Attacked."

"the most offensive patent." Argumedo quoted in "The Monopolization of Traditional Knowledge," Action Alert, Cultural Survival Canada, Jan. 15, 1997.

a blue-ribbon panel. See Committee on Genome Diversity, Commission on Life Sciences, National Research Council, *Evaluating Human Genetic Diversity* (Washington, D.C.: National Academy Press, 1997).

149 *Sequana Therapeutics.* Eliot Marshall, "Gene Prospecting in Remote Populations," *Science*, Oct. 24, 1997, p. 565.

"this is one area of genetic studies." Ibid., p. 565.

9. THE LANDSCAPE OF INVENTION

154 *Those heralding the Third Wave.* Alvin and Heidi Toffler, *The Third Wave* (New York: Bantam Books, 1981).

155 *"To be sure, there are still."* Walter Kiechel III, "The Value of the Intangible," *Fortune*, Oct. 3, 1994, p. 6.

156 *National Inventors Expo.* The other sponsor of the annual National Inventors Expo is a nonprofit organization, Intellectual Property Owners. The specific participants and patents mentioned were exhibited at the National Inventors Expo, July 23, 1994.

157 *"When I walk into my office."* Lehman quoted in Seth Shulman, "Patent Medicine," *Technology Review*, Nov.–Dec. 1995, pp. 29–36. See also Bruce A. Lehman,

157 "Intellectual Property: America's Competitive Advantage in the 21st Century," *Columbia Journal of World Business*, Spring 1996, pp. 6–17.

157 *"The knowledge-based economy does not behave."* Peter F. Drucker, *Post-Capitalist Society* (New York: Harper Business, 1993), pp. 183–84.

One such biotechnology patent. See U.S. Patent No. 4,237,224, "Process for Producing Biologically Functional Molecular Chimeras," issued to Stanley N. Cohen and Herbert W. Boyer, Dec. 2, 1980, and assigned to Stanford University. The earnings of $150 million are reported in Steven Benowitz, "Is Corporate Research Funding Leading to Secrecy in Science?" *The Scientist*, Apr. 1, 1996, p. 3.

158 *Edison's laboratory in West Orange.* Unless otherwise noted, information derives from my visit to the Edison National Historic Site, Sept. 1996.

159 *"best equipped and largest Laboratory."* Elfun Society, Algonquin Chapter, *The Edison Era, 1876–1899* (Schenectady, N.Y.: Elfun Hall of History, General Electric Research and Development Center, 1976).

a newspaper report from 1887. Oliver E. Allen, "The Power of Patents," in Frederick Allen, ed., *Great Inventions That Changed the World*, special issue of *American Heritage of Invention and Technology*, 1994, pp. 2–10.

three thousand separate materials. Robert Friedel, "New Light on Edison's Light," in Allen, *Great Inventions That Changed the World*, pp. 10–14.

160 *"genius is 1 percent inspiration."* Kathleen McAuliffe, "The Undiscovered World of Thomas Edison," *Atlantic Monthly*, Dec. 1995, pp. 80–93.

"always tying the 'R' with the 'D.'" Interview with Gregory Field, Nov. 1996.

161 *"When the twentieth century rolled around."* Peter Lynch and John Rothchild, *Learn to Earn* (New York: Fireside, 1995), pp. 60–70.

a kind of "infostructure." I am indebted to Marc Miller for stressing the importance of this particular concept.

163 *"de-massify" society.* Esther Dyson, George Gilder, Jay Keyworth, and Alvin Toffler, "A Magna Carta for the Knowledge Age," *New Perspectives Quarterly*, Fall 1994.

$1 trillion in corporate mergers. Leslie Wayne, "Wave of Mergers Is Recasting Face of Business in U.S.," *New York Times*, Jan. 19, 1998, p. 1.

consolidation already approaches the limit. See, for instance, Leslie Wayne, "800-Pound Guests at the Pentagon," *New York Times*, Mar. 15, 1998, p. 5; and Edmund L. Andrews, "Hansel & Gretel, Inc., Bertelsmann's Journey from Fairy Tales to Multimedia," *New York Times*, Mar. 24, 1998, p. D1.

10. THE NEW MONOPOLIES

165 *Jerome Lemelson.* Unless otherwise cited, material on Jerome Lemelson derives from interviews with him in March and April 1995 and from materials provided by his public relations staff at Lipman Hearne. See also Robert Thomas, Jr., "Jerome H. Lemelson, an Inventor, Dies at 74," *New York Times*, Oct. 4, 1997, p. A16. See also John Flinn, "Father of Invention," *San Francisco Examiner*, Apr. 9, 1995, p. B1.

166 *"a thinker, not a tinkerer."* Interview with Anne Cunniff, Mar. 1995. See also Seth

Shulman, "Are You a Problem Solver?" (Profile of Jerome Lemelson), *Parade,* May 6, 1996.

166 *Velcro-covered Ping-Pong balls.* Lawrence D. Maloney, "Engineer of the Year: Jerome Lemelson," *Design News,* Mar. 6, 1995, p. 75.

167 *U.S. Patent No. 5,351,078.* "Apparatus and Methods for Automated Observation of Objects," issued to Jerome Lemelson, 1994. Discussion of the patent draws upon Greg Aharonian, *Patnews,* Internet patent news service, Oct. 5, 1994.

"ink-blot tests." Robert Bell, "What Is a Submarine Patent? Who Is Jerry Lemelson?" on-line discussion, Nov. 14, 1996, at http://ftplaw.wuacc.edu/listproc/.

168 *"If you file often enough."* Ibid.

U.S. Patent No. 5,707,114. The patent was issued to Raphael Schlanger, Jan. 1998.

169 *"True, a few inventors may watch."* See "The Harm of Patents," *The Economist,* Aug. 22, 1992, p. 17.

"I could file a patent on a red chair." William Budinger, "Talk of the Nation," National Public Radio, Feb. 3, 1998 (broadcast transcript by Federal Document Clearing House, Inc.).

170 *"poetry can only be made."* Frye quoted in James Boyle, *Shamans, Software, and Spleens: Law and the Construction of the Information Society* (Cambridge, Mass.: Harvard University Press, 1996), p. 57.

"All new knowledge." Bruce Hartford, "Writers on the Information Plantation," panel presentation at the Computer Professionals for Social Responsibility Conference, Oct. 1997, reprinted in *CPSR Newsletter,* Fall 1997, pp. 12–15.

171 *he won more than $500 million.* Thomas, "Jerome H. Lemelson, an Inventor, Dies at 74."

as Lemelson confided. Interview with Jerome Lemelson, Mar. 1995.

Lemelson won $71 million. Charles J. Murray, "America's Master Inventor," *Design News,* June 3, 1991.

widely hailed as a growth area. See Thomas G. Field, Jr., "Mining the Information Age," background on-line information compiled by the Franklin Pierce Law Center, Durham, N.H., updated Apr. 1997. Available at http://www.fplc.edu.

patent litigation in the field of biotechnology. "Biotechnology Patent Litigation up 69% from 1995, According to AIPLA Survey," press release, Morgan & Finneran (intellectual property law firm), New York, Feb. 27, 1997. See also H. Lee Murphy, "Patently Offended: Lawsuits Rise over Corporate Cribbings," *Crain's Chicago Business,* Apr. 14, 1997.

172 *Digital Equipment Corporation* Joann Miller, "Digital's Patent Suit: Intel not Inside," *Boston Globe,* May 14, 1997, p. D1.

Sun Microsystems. John Markoff, "Sun Sues Microsoft on Use of Java System," *New York Times,* Oct. 8, 1997, p. D5.

Hoffman-La Roche. Marcia Barinaga, "Scientists Named in PCR Suit," *Science,* June 2, 1995, p. 1273.

now costs litigants $1.2 million. American Intellectual Property Law Association,

"Table 22. Estimated Costs of Litigation by Location of Primary Place of Work," *AIPLA 1997 Economic Survey* (Arlington, Va.: AIPLA, 1998), p. 66.

173 *even a preliminary court skirmish.* Greg Aharonian, *Patnews,* Internet patent news service, July 5, 1997.

"In the struggle to compete." Fred Warshofsky, *Patent Wars: The Battle to Own the World's Technology* (New York: John Wiley and Sons, 1994), p. 9.

the case of Dow Chemical Company. Polly LaBarre, "The Rush on Knowledge," *Industry Week,* Feb. 19, 1996, p. 53.

"the new global currency of technology." "SmartPatents Delivers Business Decision System for Intellectual Property," press release, *Business Wire,* June 16, 1997.

174 *intriguing antitrust case.* James Gleick, "Making Microsoft Safe for Capitalism," *New York Times Magazine,* Nov. 5, 1995. See also Kevin J. Arquit, "Cracking Down on Microsoft: Neglecting Anticompetitive Behavior in the Computer Industry Could Prove Catastrophic," *Intellectual Property,* Jan. 1998.

Windows controls 86 percent. See "Microfuture," *Business Week,* Jan. 19, 1998, p. 61.

U.S. district court issued an injunction. Steve Lohr, "Smokestack Doctrine, Digital Age," *New York Times,* Feb. 15, 1998, p. 5.

there is no fundamental difference. Robert Kuttner, "Bill Gates, Robber Baron," *Business Week,* Jan. 19, 1998.

a new economic view. Joe Sims and Jeffrey A. Levee, "An Antitrust Harmonic Convergence," *Intellectual Property,* Jan. 1998.

175 *"network effects."* Ibid.

"There is still an honest debate." Rubinfeld quoted in John Cassidy, "The Force of an Idea," *New Yorker,* Jan. 12, 1998, p. 32.

11. THE MOST PRECIOUS ASSET

178 *roughly one hundred thousand genes.* For a detailed discussion of what the project entails, see Nicholas Wade, "The Struggle to Decipher Human Genes," *New York Times,* Mar. 10, 1998, p. C1.

179 *"Superficial slogans."* George Poste, "The Case for Genomic Patenting," *Nature,* Dec. 7, 1995, p. 534.

"Contrary to the claims." Jonathan King, "Gene Patents Retard the Protection of Human Health," *GeneWatch, Bulletin of the Council for Responsible Genetics,* Oct. 1996, pp. 10–11. See also Declan Butler, "Agreement Urged on Access to DNA Databases," *Nature,* July 7, 1994, p. 4. And see Sandi Dolbee and Craig D. Rose, "Gene Patents Come under Fire," *San Diego Union-Tribune,* May 18, 1995, p. A1.

180 *some eight thousand patents.* Interview with Robert Cooke-Deegan, Mar. 1998. In a project begun at the U.S. Office of Technology Assessment (now defunct), Cooke-Deegan has personally reviewed all patents issued to date pertaining to human genes.

180 *one scientific team made headlines.* David Dickson, "Open Access to Sequence Data Will Boost Hunt for Breast Cancer Gene," *Nature,* Nov. 30, 1995, p. 425.

at least 15 percent of the information. David Brown and Rick Weiss, "Scientists Glimpse Genes' Division of Labor," *Washington Post,* Sept. 28, 1995.

"It will be an enormous ball and chain." Sulston quoted in Dickson, "Open Access," p. 425.

181 *nearly $3 billion.* See, for example, Michael Gruber, "Map the Genome, Hack the Genome," *Wired,* Oct. 1997, p. 153.

genome map "will be the topic." Collins quoted in Sue Goetinck, "Gene Whiz!" *Dallas Morning News,* Oct. 16, 1995, p. 6D.

Begun in 1969. For a detailed account, see Kat Hafner and Matthew Lyon, *Where Wizards Stay Up Late: The Origins of the Internet* (New York: Simon and Schuster, 1996), chapters 4 and 6.

"most universal and indispensable network." As quoted in Jon Wiener, "Free Speech on the Internet," *The Nation,* June 13, 1994, p. 825.

182 *at least 57 million Americans. New York Times,* Apr. 9, 1998, p. D1.

the number of Web pages. Statistics compiled by Relevant Knowledge, Inc., Atlanta, Ga.

one of the ten worst ideas. Paul Magnusson, "The 10 Worst Ideas of the Campaign," *Business Week,* Nov. 11, 1996, p. 39.

183 *"To fashion a subsidy system."* Harter quoted in John Simons, "There's No Free Lunch in Cyberspace: Clinton's Plan to Wire Schools Runs Into a Wall," *US News and World Report,* Dec. 9, 1996, p. 72.

National Library of Medicine. See "Library of Medicine Database Opened to the Public," press release, National Institutes of Health, June 30, 1997. For MedLine access see: http://www.ncbi.nim.nh.gov/PubMed.

Project Gutenberg. See "First Large, Public Virtual Library," in Brad Wieners and David Pescovitz, *Reality Check* (New York: Wired, 1996), p. 97.

"Without explicit, vigilant attention." Michael Dertouzos in "Roundtable Discussion of Internet Issues," *Technology Review,* July 1996. Participants included Mitchell Kapor and Amy Bruckman.

184 *like Herman Daly's invisible foot.* As described in Chapter 3. See Herman Daly, "Towards a Stationary-State Economy," in John Harte and Robert Socolow, eds., *Patient Earth* (New York: Holt, Rinehart and Winston, 1971).

185 *Nathaniel Pitt Langford.* See Nathaniel Pitt Langford, *The Discovery of Yellowstone Park: Journal of the Washburn Expedition to the Yellowstone and Firehole Rivers in the Year 1870* (Nebraska: University of Nebraska Press, 1972). Langford went on to be the first superintendent of Yellowstone National Park.

a matter of continued dispute among historians. For a detailed discussion of the founding of Yellowstone, see Paul Schullery, *Searching for Yellowstone: Ecology and Wonder in the Last Wilderness* (Boston: Houghton Mifflin, 1997), pp. 51–67.

Yellowstone Park Act in 1872. From "Yellowstone Park History" Web site, copyright Bruce T. Gourley, 1996. See http://www.yellowstone.com/.

dismissed as a failure. Ibid.

185 *John Muir.* For more on this fascinating, radical environmentalist, see Stephen Fox, *John Muir and His Legacy: The American Conservation Movement* (Boston: Little, Brown, 1981).

185 *"Lying out at night."* Roosevelt quoted in Kevin J. Downing, "The Man Who Saved Yosemite," *Scholastic Update,* Apr. 15, 1994, p. 15.
tripled the amount of land. Ibid.

186 *George W. Field's appreciation.* John H. Mitchell, "Forever Wild," *Sanctuary: The Journal of the Massachusetts Audubon Society,* Nov.–Dec. 1997, p. 1.
80 million acres. See Edward Hoagland, "The Best Idea," *Life,* special issue, *Our National Parks: A 75th Anniversary Celebration,* Summer 1991, p. 9.
3 million visitors annually. Sherri Byrand and Karl Byrand, "Grace under Pressure," *National Parks,* Mar.–Apr. 1997, pp. 27–30.
140 nations have emulated. See "Yellowstone Park History" Web site.

187 *Andrew Carnegie derided.* As reported in the *San Francisco Chronicle* about 1910. Reprinted in *John Muir Newsletter* 1, no. 2 (Spring 1991).

188 *"These parks did not just happen."* See Hoagland, "The Best Idea," p. 9.
our conception of ownership. Vivian Weil and John W. Snapper, eds., Introduction to *Owning Scientific and Technical Information: Value and Ethical Issues* (New Brunswick, N.J.: Rutgers University Press, 1989), pp. 169–92.

191 *Universal Declaration of Human Rights.* United Nations, "Universal Declaration of Human Rights," adopted December 10, 1948.

Acknowledgments

Perhaps it is the subject matter, the time spent thinking about the debts we all owe for our creative work. Or perhaps it is my background as a journalist, with reporting deriving more exclusively from firsthand accounts of people, places, and events. But, whatever the reasons, I feel that I owe a great intellectual debt to many other thinkers, writers, and journalists for this book. Due to the breadth and nature of the topic, I have drawn heavily on all sorts of primary and secondary materials — from historical accounts to legal analyses, from obscure local newspaper clippings to esoteric on-line postings. I have tried to faithfully cite these contributions in the notes. My work is woven from threads often painstakingly gathered by this wide-ranging group, and I am grateful for the chance to incorporate their accounts and build upon their insights.

On a more personal note, this book was many years in the making, and my thinking was aided immeasurably by discussions and substantive help from many quarters. Marc Miller, my longtime friend and former editor, conceived of the idea with me and spent many long months in discussions that helped me shape the initial work. He sent a steady stream of material across my desk and provided insightful comments on early versions. I am very grateful for his help and friendship.

Many others lent encouragement and insight to the project at key junctures along the way. I especially thank Barbara Goldoftas, Sandra Hackman, Ken Kimmell, and Alan Cooperman for their moral support and good counsel. Steve Marcus and Phil LoPiccolo, formely at *Technology Review,* helped me grapple with some of this material in articles for that magazine. The section on Jerome Lemelson is based in part on an assignment for *Parade.*

A far larger contingent helped set me straight on tricky points of law, science, medicine, or economics, either in written comments, in-depth interviews, or particularly good dinner conversations. Among them, Mario Aieta, Jim Rebitzer, Hope Shand, Bennett Goldberg, Phil Bereano, Albert Butzel, and Richard Stallman all offered insights that contributed in some measure to my grasp of the material. Of course, any errors or misinterpretations are mine alone.

Hearty thanks as well to the team at Houghton Mifflin, especially Steve Fraser and Eric Chinski for their incisive editorial help, Wendy Strothman for her confidence in the project, and Peg Anderson for her skilled and careful copy editing. I am grateful as well to my agent, Katinka Matson, at Brockman, Inc., and to my former editor Deanne

Urmy at Beacon Press for her support early on and continued friendship. I offer a special mention of appreciation for all the encouragement and support given by my maternal grandmother, Lena Wolf.

Most of all, though, there is simply no way to adequately thank my unbeatable sustenance network on the home front — Elise, Ben, Roy, Karen, Sarah, Jill, Tom, Zoe, and, most of all, Laura — who did so much on every level to make this project possible.

Index